State and Local Issues in Transportation of Hazardous Waste Materials:
Towards a National Strategy

Proceedings of the National Conference on
Hazardous Materials Transportation

sponsored by the
Urban Transportation Division
of the
American Society of Civil Engineers

in cooperation with the
Transportation Research Board
and
U.S. Department of Transportation

St. Louis, Missouri
May 14-16, 1990

Co-edited by Mark D. Abkowitz, Vanderbilt University
and
Kostas G. Zografos, University of Miami

Published by the
American Society of Civil Engineers
345 East 47th Street
New York, New York 10017-2398

ABSTRACT

This proceedings contains papers presented at the specialty conference State and Local Issues in Transportation of Hazardous Materials: Towards a National Strategy held on May 14-16, 1990 in St. Louis, Missouri. The papers span a wide spectrum of technical, regulatory, and policy aspects of hazardous materials transportation. The papers are organized into the following five thematic areas: 1) legislative issues, 2) risk perception, 3) database development, 4) risk assessment/routing, and 5) risk management and advanced technologies. Within these broad categories, such topics as the need for local control, the design of protective systems into highways systems, the use of information to assess accident trends, the importance of individual and societal risk analysis in decision-making, and the development of computer software to assist in routing hazardous materials are explored.

Library of Congress Cataloging-in-Publication Data

National Conference on Hazardous Materials Transportation (1990: Saint Louis, Mo.)
   State and local issues in transportation of hazardous waste materials: towards a national strategy: proceedings of the National Conference on Hazardous Materials Transportation: St. Louis, Missouri, May 14-16, 1990/sponsored by the Urban Transportation Division of the American Society of Civil Engineers in cooperation with the Transportation Research Board and U.S. Department of Transportation: co-edited by Mark D. Abkowitz and Kostas G. Zografos.
   p.   cm.
   Includes indexes.
   ISBN 0-87262-796-9
   1. Hazardous substances—United States—Transportation—Congresses. I. Abkowitz, Mark David. II. Zografos, Kostas G. III. American Society of Civil Engineers. Urban Transportation Division. IV. National Research Council (U.S.). Transportation Research Board. V. United States. Dept. of Transportation. VI. Title.
T55.3.H3N275   1990
363.17—dc20   91-18477
                                                                                    CIP

The Society is not responsible for any statements made or opinions expressed in its publications.

Authorization to photocopy material for internal or personal use under circumstances not falling within the fair use provisions of the Copyright Act is granted by ASCE to libraries and other users registered with the Copyright Clearance Center (CCC) Transactional Reporting Service, provided that the base fee of $1.00 per article plus $.15 per page is paid directly to CCC, 27 Congress Street, Salem, MA 01970. The identification for ASCE Books is 0-87262/88. $1 + .15. Requests for special permission or bulk copying should be addressed to Reprints/Permissions Department.

Copyright © 1991 by the American Society of Civil Engineers,
All Rights Reserved.
Library of Congress Catalog Card No: 91-18477
ISBN 0-87262-796-9
Manufactured in the United States of America.

Foreword

The transportation of hazardous materials is inherent in any advanced and technologically complex society. A number of industrial processes of vital economic importance are dependent on the uninterrupted flow of hazardous materials shipments. Although the production of hazardous materials is associated with technological growth and economic development, there is significant potential danger to the population and the environment in the event of a release. The risks associated with the transportation of hazardous materials have drawn considerable attention at local, national and international level, resulting in the development of a regulatory framework to enhance the safety of hazardous materials movements.

Many responsibilities associated with the safe transport of hazardous materials fall to state and local goverenments. Such tasks include community preparedness and emergency response, risk evaluation and communication, routing and siting considerations, data collection and information management, vehicle inspection and regulation en- forcement. The extent of assumption of these responsibilities varies across the country, and jurisdictional purview in some areas is subject to debate. Nevertheless, several exemplary activities have been undertaken by state and local governments, which are compatible with the development of a national strategy and which have been beneficial in managing the safe transport of dangerous goods.

The principal objective of this conference was to enchance the exchange of information concerning effective state and local methods for managing hazardous materials transportation within an evolving national system. To achieve the conference objectives, papers reflecting a wide spectrum of local and state issues in hazardous materials were included in the program.

This volume includes copies of the papers presented at the conference. The papers are organized into the following thematical categories: 1) legislative issues, 2) risk perception, 3) database development 4) risk assessment and routing, and 5) risk management and advanced technologies.

The importance of the issues covered by the conference is underscored by the extensive list of participating organizations who banded together to structure the technical program:

    American Association of State Highway and
    Transportation Officials
    American Petroleum Institute
    American Trucking Associations
    Association of American Railroads
    Hazardous Materials Advisory Council
    International Bridge, and Tunnel and Turnpike Association
    National Association of Governors' Highway Safety Representatives
    National Association of Regulatory Utility Commissions
    National Association of Towns and Townships
    National Conference of State Legislatures
    National League of Cities
    National Safety Council
    National Tank and Truck Carriers Association
    U.S. Conference of Mayors
    U.S. Department of Energy

The conference participation was impressive not only in terms of the number of participants attending the three days of sessions and exhibits, but also in terms of the diversity of their technical and professional background.

Several individuals made notable contributions to the planning and implementation of the conference on "State and Local Issues in Transportation of Hazardous Materials: Towards a National Strategy." They are cited according to their position of responsibility associated with the conference:

Conference Organizing Committee

Mark Abkowitz, Chairman
Ashok Boghani, Vice-Chairman
Paula Alford
James Brogan
Michael Bronzini
Joel Brown
John Cashwell
Robert Chipkevich
Sherwood Chu
Michael Demetsky

Richard Doyle
T. Duncan Ellison
George List
Claire Orth
Edith Page
Essam Radwan
William Rhyne
Paul Rothberg
Eugene Russell

Conference Steering Committee

Susan Armstrong
Paul Bomgardner
Judith Burrell
Francis Francois
Mauren Gallagher
Stacy Gerard
Barbara Harsha
Clifford Harvison
Charles Keller

Frank Lisle
Margaret Matheson
Paul Rankin
James Reed
Paul Rogers
Jeffrey Shiff
Richard Tippie
Nick Yaksich

The conference was sponsored by the: 1) American Society of Civil Engineers, 2) Transportation Research Board, and 3) U.S. Department of Transportation.

The American Society of Civil Engineers served as the host organization in handling the logistics of conference planning and publication of these proceedings. Shiela Menaker and Elizabeth Yee of ASCE were primarily responsible for overseeing conference planning and publication activities.

The review process in approving papers for publication consisted primarily of assuring conformity with publication guidelines issued by ASCE to the authors. A paper in the proceedings will not be eligible for other publication by ASCE unless the Urban Transportation Division within ASCE approves it for republication or unless the paper has been significantly revised or expanded. The papers in the proceedings will be eligible for ASCE prizes and awards under the rules that govern such awards, and will be open to discussion in the ASCE Journal of Transportation Engineering.

Orders of additional copies of State and Local Issues in Transportation of Hazardous Materials: Towards a National Strategy, may be placed with the Publications Fulfillment Department, c/o ASCE Headquarters, 345 East 47th Street, New York, NY 10017-2398.

TABLE OF CONTENTS

Keynote Address: T. Dungan . . . . *

I. HAZARDOUS MATERIALS LEGISLATIVE ISSUES

Local Government Views: Hazardous Materials Transport Legislation
    by T.L. Novak 2
State Legislative Concerns Relative to Federal Hazardous
Materials Regulation
    by J.B. Reed 9
Dangerous Goods Emergency Response: The Western Australian
Experience
    by K. Price 17

II. HAZARDOUS MATERIALS RISK PERCEPTION

Risk Management in the Transportation of Dangerous Goods
    by M.K. Matthews 34
Developing High-Risk Scenarios and Countermeasure Ideas For
Mitigation of Hazardous Materials Incidents
    by E.R. Russell, Sr. 38
Risk Perception and Policy Preferences of Fire Chiefs: A Survey of
Hazardous Materials Issues
    by S. Warkov, and K.G. Zografos 52

III. DEVELOPMENT OF HAZARDOUS MATERIALS DATABASES

Hazardous Materials Data: A Federal Perspective
    by R.C. Hannon 62
Flows of Hazardous Materials Through States by Rail
    by R.C. Hannon and Paul Zebe 74
Issues in Developing Arizona's Hazmat Incident Reporting System
    by K.S. Hubbard 87
Developing a HazMat Incident Evaluation Program
    by D.L. Williams and F. Kaikumba 94
Canadian Database Development as a Support Tool to Transport
Risk Assessment
    by D.A. Learning 104

* Manuscript not available at time of printing

IV. HAZARDOUS MATERIALS RISK ASSESSMENT AND ROUTING

Managing Risks of Hazardous Materials Transportation in
Santa Barbara County
    by D.K. Anthony and J. Peirson 120
Assessing Community Safety for Hazardous Materials Transport
    by C.K. Chiang, E.J. Cantilli and S.T. Ying 134
Highway Robbery: The Social Costs of Hazardous Materials
Incidents on the Capital Beltway
    by T.S. Glickman and M.K. Macouley 148
A Community-Focused Routing and Siting Model for Hazardous
Materials and Wastes
    by G. List and P. Mirchandani 163
Truck Accident and Release Rates for Hazmat Routing
    by D.W. Harwood, J.G. Viner and E. R. Russell, Sr. 177
Evaluation of Hazardous Materials Transportation by Rail
    by W.H. Oderwald and M.A. Sontag 191
A Risk Vulnerability Assessment Approach for Selecting Routes:
Case Study of Hazardous Waste Transport in Arizona
    by K.D. Pijawka, A.E. Radwan and J.A. Soesilo 205
Societal-Individual Risks for Hazmat Transport
    by F.F. Saccomanno and J. H. Shortreed 220

V. HAZARDOUS MATERIALS AND ADVANCED TECHNOLOGIES

GIS Risk Analysis of Hazardous Materials Transport
    by C. Anders and J. Olsten 248
Use of Advanced Technologies for Improving Hazmat Transportation
Safety
    by A. Boghani 262
StateGEN/StateNET and DOT Guidelines: Tools for Highway Routing
of Hazardous Materials
    by J.W. Cashwell, J.D. Brogan and C.M. Erickson 271
Computer-Assisted Risk Assessment of Dangerous Goods
Transportation for Haute-Normandie
    by K. Fedra, S. Lassore and E. Weigkricht 281

Subject Index 295
Author Index 297

LEGISLATIVE ISSUES

LOCAL GOVERNMENT VIEWS

HAZARDOUS MATERIALS TRANSPORT LEGISLATION

Terry L. Novak, Ph.D*

This paper presents the views of American local governments. This may appear presumptuous because of the great diversity between cities and counties, regional differences, and the differences in size, but I have relied upon the policy guidance of the National League of Cities and these comments have been reviewed by a number of local officials with experience in this issue. The thoughts here are presented from the perspective of the Mayor and City Council, City Manager, and Police and Fire Chiefs of the urban communities who must deal with transport accidents. We attach as an exhibit the policy position of the National League of Cities.

As a general rule, as we have discovered in the application of SARA Title III, while some of the large cities and the large industrial and transport firms are in compliance with new statutes and can be expected to do a professional job, most cities, rural areas with rural fire departments, and firms without professional industrial safety personnel are not currently in compliance. They are usually the ones who are slow in responding to statutory provisions and need the most training and equipment. In many instances, staffing must be increased to administer pre-emergency regulations applying to both local government and businesses and to ensure that emergency response teams can safely and effectively carry out their missions. Federal mandates should also provide for federal support.

Values and Attitudes

Local public safety agencies obviously do not operate in a vacuum. We all have close relationships with the State Patrol, the State Environmental Protection Agency, and very often are tied together into regional hazardous materials teams. Nonetheless, there are certain values and attitudes found among local government officials which may not be shared by other participants in this legislative process.

City officials, whether elected or appointed, are readily accessible by citizens and have become media sensitive. Hazardous materials accidents are a "hot button" for the media, and citizens expect that their elected and appointed officials are on top of these situations. There is nothing so frustrating and productive of anger as a local elected official being questioned by a television station and having either no information about an incident or wrong information. We are also skeptical of the state and federal bureaucracies.

Hardly a meeting of local officials goes by in which someone does not offer anecdotes about state and federal impractical responses which obviously would not have occurred had the locals been in charge. We also are concerned, both in a legalistic and in a humane sense, about danger to our employees, whether they be fire fighters, police officers, or public works employees responding to such incidents. This is one of the concerns we share with the transport industry. Finally, hardly a week seems to pass without a hazardous materials accident somewhere in North America being reported in the media. We are worried.

Historically, local governments are concerned about doing a professional job of emergency response. Most communities place great pride in their police and fire departments. While fighting actual fires is today a reduced proportion of the usual fire department's efforts, emergency response to hazardous materials incidents and medical calls are dramatically increasing. In addition, we naturally feel that state and federal regulations should be a floor, not a ceiling. The diversity of urban locales is so great that we view with derision any attempt to govern from afar.

There is, however, a recognition of the need for joint action and some sort of exemptions. We are not intent upon breaching national security and if a valid argument can be made that shipments by the Department of Energy and Department of Defense should be exempt, so be it. (We do not, however, see a relationship between hazardous shipments and national security.) We also see the need for a dispute resolution procedure (with court access) applied to disputes in this arena. Finally, we see the need for uniform training requirements, uniform national registration, a uniform inspection format, standard container regulations, and many other uniform provisions in this nation-wide industry. In fact, the National League of Cities' policy statement cites nine different items in which we support uniformity.

In summary, while the local governments of this nation have historical concerns and skepticism about both the industry and state and federal bureaucracies, we certainly are not blind to the need for uniformity, dispute resolution procedures, and exemptions.

Legislative Suggestions

As the legislation that we're discussing at this conference moves forward:

1) We seek a universal registration fee for trucking firms involved in this industry with federal standards for training, inspection, and licensing, but a licensing arrangement operated by the states.

2) These registration fees would build a trust fund for training and equipment of hazardous materials response teams, with those funds distributed among the various localities in some proportion to the degree of risk their populations face from such transportation. Congress' Office of Technology Assessment has estimated that only 25 percent of the approximately 2 million people in the emergency response network have adequate training to meet a hazardous materials emergency.

3) Within federal performance standards, we feel a joint decision-making process on routing questions can be made to work, given a national study of routes and cargos made available to all parties. The local governments of the nation will themselves undoubtedly be part of the problem in this instance. Central cities will wish to route hazardous materials transportation around the central core through the suburbs, who of course will protest vehemently. Nonetheless, we feel with proper mediation and arbitration arrangements a joint decision-making arrangement can be made to work.

4) We do not want to be notified of every truck and train shipment of hazardous cargo. Such data would be so voluminous as to be useless; but we do wish notification to local responders of extremely hazardous cargo so we can be prepared. The development of a centralized tracking system for all en-route shipments of hazardous materials would allow emergency responders the use of another informational base during those times when the bill of lading, placards, or shipping documents cannot be accessed. The inability to timely identify the products involved is one of the major problems of emergency responders. Major transportation routes can be shut down unnecessarily and public safety personnel and equipment allocated for great lengths of time for an innocuous spill that could not be ruled out as a hazardous materials incident.

5) We feel the federal government could develop national container regulations based on realistic functional standards which those firms who manufacture containers for trucking and railroads could meet.

6) We are not convinced that the federal safety regulators have seriously policed this industry, either trucking or rail. We need a national system to share information on trucking firm safety ratings plus a major infusion of federal funds to the federal inspection service.

7) We all need to turn our attention to reducing the volume of such wastes. The EPA's proposed re-write of the Resource Conservation and Recovery Act will supposedly pursue that goal.

8) Finally, we get to the core of the problem, and that is flexibility for the states to pass regulations which go beyond the federal regulations, and for localities, if necessary, to go beyond the states, but we recognize that we must be prepared to show a special need in those instances. The geographic, demographic, and climatic diversity of this nation is so great that a simple federal standard applied inflexibly across from California to Maine makes no sense to us. Those who respond to these incidents are not members of the United States Fire Department -- there is no such agency.

Political Environment

Finally, let me comment on the political environment and offer some conclusions. The public is learning to care deeply about this issue. It has become a major item in the media, and opinion polls show citizens throughout the nation are becoming concerned about hazardous chemicals in both transport and stationary use. We recognize that those who transport this material by truck or train have valid concerns about a patchwork of non-uniform regulations and overzealous enforcement. While our interests sometimes diverge, they can also converge, because we share with the industry our concern for employee safety. A truck driver injured by hazardous chemicals and a fire fighter injured in his spill both go to the same hospital, perhaps in the same ambulance. In addition, lawsuits and high insurance rates plague all of us. We are optimistic that modern computer technology can ease the data dissemination problem and, except in cases of true national security, see no reason why extremely hazardous cargo information and general information about trends in the industry cannot be shared. Finally, I am sure most of us at this conference agree that the sooner we move forward on this legislation the better for all.

G. Transportation of Hazardous Materials

1. Consolidation of Federal Programs

Overlapping and conflicting hazardous materials regulatory policies should be eliminated and federal agencies involved in the regulation of hazardous materials (such as DOT, EPA, DOE, FEMA and NRC) should coordinate their programs much more closely. As a first step, the Department of Transportation, the Environmental Protection Agency and other appropriate federal agencies should consolidate their lists of hazardous materials, wastes and substances into a single comprehensive listing. In those instances where more than one federal agency regulates a hazardous material, waste or substance, the appropriate agencies should examine existing regulations and take steps to eliminate conflicting or inconsistent policies.

2. Container Standards

The federal government should set stricter standards for hazardous wastes, hazardous materials and radioactive materials containers. Standards which are design specifications should be replaced with functional performance standards which would give manufacturers more latitude in the development of safer containers.

Additionally, rail tank car designs should come under closer federal scrutiny. The Research and Special Program Administration (RSPA) and the Federal Railroad Administration (FRA) should serve as consultants on the Association of American Railroads' (AAR) Tank Car Committee (the Committee which develops and adopts industry standards upon which federal design specifications are based) when there is a specific tank car design problem that is being addressed by that Committee.

3. Data Collection

Information about the general nature and quantity of materials being shipped, the general shipment route, and other relevant information about the shipper, including, without limitation, shipments made by or under the direction of the United States Departments of Energy or Defense, should be collected on an annual basis and disseminated in a manner to be determined by the Secretary of Transportation.

4. Routing

The federal government should develop performance standards to guide the selection of highway and rail routes along which hazardous materials can be shipped, including, without limitation, shipments made by or under the direction of the United States Department of Energy or Defense. State and localities should have concurrent jurisdiction over routing; that is they should be allowed to adopt and enforce highway and rail routing requirements (including time-of-day restrictions, escorting, and local bans) which are consistent with the federal performance standards, unless the state or locality can demonstrate that it has a unique local safety or health circumstance which necessitates passage of a routing requirement inconsistent with the federal standard. To the extent possible, the federal government should encourage the use of public/private sector coordinating councils, composed of representatives of industry and state and local governments, to promulgate routing requirements and select appropriate routes. The federal government should provide technical assistance to localities so that these jurisdictions can make safer, more informed routing decisions.

The federal government should enforce preemption of inconsistent state and local highway and rail regulations, as required under current law, except for those regulations developed in accordance with the federal routing standards.

Disputes about the consistency of a routing requirement with federal law should be considered by the Department of Transportation. DOT should evaluate inconsistency petitions according to specific criteria (such as the impact of the requirement on interstate commerce and safety) and there should be time limits on the evaluation in order to expedite decision-making. In a court review, the Secretary of Transportation's decision should be advisory, not binding upon the court.

5. Frequency

Maximum effort should be made by the federal government to minimize the number of high level radioactive material shipments, including, without limitation, shipments made by or under the direction of the United States Department of Energy or Defense. The federal government should require a showing of need wherein the shipper must show that the shipment is clearly preferable to other alternatives.

6. National Hazardous Materials Driver Standards

To improve the safety of truck transportation of hazardous materials, the federal government should create uniform national truck licensing and training standards for drivers of interstate and intrastate vehicles transporting hazardous materials. As a prerequisite to a license, drivers of such vehicles should be required to certify that they have received special training on hazardous materials handling, routing, and emergency response. Training and licensing standards should be issued by the U.S. Department of Transportation and administered by the states.

Under this approach, states would develop a licensing program that is at least as stringent as the federal standards, administer the training requirements, test potential licensees, collect the licensure fees, enforce the federal standards and exchange license information (in order to prevent multiple licensing) and accident and violations information (in order to create a centralized drivers record) with each other.

7. Local Notification

Shippers should be required to give general, not shipment by shipment notification to cities along a prospective route of shipments of hazardous materials, including, without limitation, shipments made by or under the direction of the United States Department of Energy or Defense. Cities receiving routing information from the United States Departments of Energy or Defense should treat such information with the confidentiality necessary to protect national security and eliminate any risk to the public health and safety from the disclosure of such information.

Routing information should also be made available to the local communities on an annual basis. Furthermore, federal, state, regional and local governments should cooperate more closely in the development of a unified data and notification system (such as a single manifest system) and in the exchange of shipment-related information, including, without limitation, shipments made by or under the direction of the United States Department of Energy or Defense.

Railroads should be required to make an immediate oral notification to designated local emergency personnel when a significant hazardous materials incident occurs (involving the death or injury of any person.

8. *Emergency Response*

Emergency response should continue to be the primary responsibility of state and local governments. However, several steps can be taken by all levels of government and by industry to improve their emergency response capabilities.

First, federal, state and local governments should develop one centralized emergency response authority within each of their jurisdictions to enable better interjurisdictional communications when an accident does occur.

Second, the federal government should develop an outreach program to explain its regulations to state and local officials and should set minimum training and education standards so that shippers and emergency personnel in all jurisdictions will have a common understanding of accident prevention and mitigation.

Third, the federal government should provide guidelines for planning for and responding to emergencies and should furnish technical information in an emergency situation. Further, the Federal Emergency Management Agency should offer more emergency response training courses for rail-related incidents, should coordinate existing rail-related courses, and should provide greater outreach to first responders on the availability of such courses.

Fourth, the federal government should evaluate and certify training programs used by regional, state and local emergency personnel.

And fifth, industry should be required to participate in education and training efforts, and in accident prevention and response activities. The railroads should be encouraged to jointly develop emergency response plans with local governments. The railroads should also be encouraged to offer more courses oriented toward the first responder, especially courses which would require minimal travel and registration fees. As an inducement to industry participation in emergency response efforts, the federal government should encourage its state and local counterparts to adopt good samaritan legislation limiting industry liability if they do respond to an accident.

State and local governments should continue to coordinate and manage emergency response activities, and should participate, with the assistance of industry and the federal government, in education and training efforts.

9. *Enforcement*

The federal government should continue to be responsible for the enforcement of federal manufacturing, testing, repair, and reconditioning standards for hazardous materials containers and packaging. The federal government should develop similar requirements and should enforce such requirements for shipments made by or under the direction of the United States Departments of Energy or Defense which are currently unregulated. Additionally, the federal government should continue to enforce standards for labeling, marking, placarding, designating and classifying hazardous materials. States and localities should have concurrent jurisdiction over vehicle operating requirements and standards; that is, they should be allowed to adopt and enforce these standards and requirements, including but not limited to imposition of fines or civil penalties for infractions of these federal requirements.

To assist states and local government enforcement of federal regulations governing the highway shipment of hazardous materials, Motor Carrier Safety Assistance Program (MCSAP) grants from the Highway Trust Funds should be increased and made available to those states and local governments which voluntarily agree not to levy a fee or tax on the transportation of hazardous materials.

Several steps should be also taken to strengthen enforcement of federal railroad safety regulations which effect the rail shipment of hazardous materials. The penalty provisions of the Federal Railroad Safety Act should be increased and should be imposed upon the offending individual and not just upon the railroad company. The Act should also be amended so that it clearly gives the federal and state governments concurrent authority to enforce federal rail safety regulations which relate to hazardous materials.

Additionally, the State Hazardous Material Enforcement Development (SHMED) program, which was created by the federal government to help states establish hazardous materials transportation inspection and enforcement programs, should be retained and adequately funded and broadened to include rail-related enforcement.

Finally, the Federal/State Rail Safety Participation program, which pays half a state rail safety inspector's salary, should be continued and adequately funded.

10. *Financing*

The federal government should establish a national trust fund based upon a registration fee which will generate enough revenues to pay for state and local hazardous materials-related needs (e.g., emergency response enforcement, planning, training, equipment and personnel). Funds should be distributed to state and local governments based on a formula which takes into account the amount of hazardous materials shipments passing through each jurisdiction annually, and in proportion to the amount of risk that is caused by such shipments, and should include all shipments made by or under the direction of the United States Department of Energy or Defense.

In order to be eligible for funds from the trust fund, a state should demonstrate that: it has initiated a program to develop a statewide emergency response plan, designated a lead agency for administering emergency response activities; agreed to assist local governments in emergency planning and response; agreed to pass a portion of its hazardous materials entitlement to local governments for planning and emergency response purposes; and established an emergency response coordinating council consisting of state, local government and industry representatives.

Any state or locality which chooses to accept funds from the hazardous materials trust fund would be precluded from levying its own fees on the transportation of hazardous materials. A locality which can demonstrate that it has a unique hazardous materials problem or need, which is not being satisfied by its federal trust fund suballocation should, be allowed to levy its own fees on the transportation of hazardous materials.

11. Insurance

The federal government should set minimum insurance requirements for all large quantity or Type B (health hazard) radioactive materials shipments and shipments made by or under the direction of the United States Departments of Energy or Defense, to ensure that shippers are financially able to bear emergency response and cleanup expenses.

12. Federal Administration of Hazardous Materials Programs

Congress should adequately fund the Materials Transportation Bureau and other related hazardous materials offices within DOT so that they can more effectively enforce the Hazardous Materials Transportation Act, more expeditiously process inconsistency petitions and promulgate regulations, more consistently collect and disseminate data and more effectively coordinate federal hazardous materials.

State Legislative Concerns Relative to Federal
Hazardous Materials Transportation Regulation

By

James B. Reed[1]

Abstract

States have increasingly asserted their authority to regulate hazardous materials transportation, in the face of inadequate and declining federal efforts. Areas of state interest include inspection and enforcement, emergency response and routing. Pending legislation to reauthorize the Hazardous Materials Transportation Act (HMTA) would strengthen the federal role in certain appropriate areas and increase funding to states for emergency response. These developments are generally welcomed as long as areas of traditional state authority are not subject to preemption.

Introduction

State legislative concerns regarding hazardous materials transportation are many. As observers of state activities know, there has been a recent flurry of legislative activity in the hazardous materials transportation area.

First, by way of background, the National Conference of State Legislatures (NCSL) is a Denver-based, nonpartisan organization of which all 50 state legislatures and the 7,500 state legislators who serve in those bodies are members. NCSL's staff of 150 provides research and information to legislators and legislative staff; provides forums for exchanges of ideas through numerous meetings and seminars; and represents the views of states and state legislators to the various agencies of the federal government. Our Washington, D.C., office carries out this lobbying function. We deal with the whole range of issues facing state legislators, including such noncontroversial topics as abortion, taxes, and nuclear

---
[1]Policy Specialist, Energy, Science and Natural Resources, National Conference of State Legislatures, 1050 17th Street, Suite 2100, Denver, Colorado 80265.

waste transportation. NCSL has been active in hazardous materials transportation issues for several years and currently has a cooperative agreement with the U.S. Department of Energy (DOE) to keep state legislators informed of the DOE activities under the Nuclear Waste Policy Act, particularly the transportation elements of that program.

NCSL's policy position on the topic of hazardous materials transportation expresses broad concern about the following four interrelated hazardous material transportation issues:

1. The complexity of existing hazardous material transportation regulations administered by various federal agencies (the Department of Transportation [DOT], the Environmental Protection Agency [EPA] and the Nuclear Regulatory Commission [NRC]), including the identification and definition of hazardous materials;

2. The lack of sufficient financial and manpower resources at the state and federal levels to ensure necessary levels of cargo inspection and enforcement of laws and regulations relating to hazardous materials transportation;

3. The lack of comprehensive and coordinated training programs and emergency response planning; and

4. The preemption of state and local laws and regulations designed to ensure adequate protection of public health and safety, in light of the federal government's lack of financial manpower resources needed to effectively enforce federal regulations. This last issue is of particular concern.

A recent report to the U.S. Senate Committee on Commerce, Science and Transportation states, "(t)here is overwhelming evidence that the Federal Highway Administration (FHWA) (an arm of DOT) had not devoted the level of attention to hazardous materials transportation that many would judge necessary." For example, even though there are over 1,200 motor carriers that are currently transporting hazardous materials that have received an unsatisfactory safety rating from the FHWA, FHWA has never used the authority provided in the Motor Carrier Safety Act of 1984 to shut down any of these motor carriers. Some of these hazardous materials motor carriers with unsatisfactory ratings have not been audited by FHWA since 1980; many more have not been visited by FHWA since the early 1980s. In fact, the number of enforcement cases recommended for prosecution against those charged with violations of federal hazardous materials transportation regulations has decreased in recent years, according to the report. Civil penalties collected for such violations fell 33 percent from fiscal year (FY) 85 to FY 87.

In the face of inadequate and declining federal enforcement, states have passed legislation to provide funding for inspection and enforcement and emergency response preparedness, among other safety items. Thirty states have assessed fees on hazardous materials shipments through their states and several have fees for radioactive shipments, among them Colorado, Illinois, New Hampshire, Ohio, Oregon, Pennsylvania, Vermont and Wyoming. In 1989, 13 legislatures considered bills to strengthen state regulation of hazardous materials transportation. In NCSL's 1990 survey of state legislatures, 19 states indicated the probability of legislation on the hazardous materials transportation issue, addressing topics like driver training, emergency response training and funding, route designation, and vehicle inspections.

A continuing obstacle to state regulation is the preemptive actions of the U.S. Department of Transportation. States are trying to construe DOT's preemptive authority more narrowly through provisions in HMTA reauthorization bills under consideration in Congress.

State Concerns

This section of the paper expands somewhat on these topics of concern to state legislators, particularly relative to emergency response.

The first is funding. States have to deal with the whole range of emergency situations, including radiological incidents, hazardous materials spills, natural disasters, and so on. Yet funding from the federal government, when it is available, comes from a specific agency for a specific reason (Federal Emergency Management Agency [FEMA], DOE, DOT, EPA, etc.). The restrictions accompanying funding often inhibit the development of comprehensive emergency preparedness at the state level. Recognition at the federal level of the need for integrated state programs would be helpful.

Another funding issue is the lack thereof. Various versions of HMTA reauthorization floating around in Congress offer various funding mechanisms and levels and this is encouraging.

A specific funding issue relative to high-level waste transportation that will be of more concern in coming years is Sec. 180(c) of the Nuclear Waste Policy Amendments Act. That section reads:

> The Secretary shall provide technical assistance and funds to States for training for public safety officials of appropriate units of local government and Indian tribes through whose jurisdiction the Secretary plans to transport spent nuclear fuel or high-level radioactive waste. Training shall cover procedures required for safe routine transportation of these materials, as well as procedures for dealing with emergency response situations. The Waste Fund shall be the source of funds for work carried out under this subsection.

There is a continuing difference of opinion between DOE and the states on when this money should become available. DOE maintains that 3-5 years prior to shipments will be sufficient, while states argue that the planning and development of training programs requires funding at an earlier date.

Another problem area is coordination. State-federal coordination is an ongoing issue and lack of federal agency coordination between themselves is a constant source of state frustration. There has been progress in clarifying roles of federal, state and local responders and states are working together to clarify how they can help each other near borders. Efforts are also underway to ensure that radiological emergency response plans are compatible with procedures already in place for other hazardous materials.

Furthermore, the states are working to promote joint state/federal participation in enhanced safety compliance and enforcement efforts as a part of the HMTA reauthorization process. Part of this involves defining appropriate roles for states which would include the authority to select routes, conduct vehicle inspections and to require permits and assess state fees.

A related issue that NCSL has been wrestling with over the years involves attempting to resolve the paradox of intense public concern about radioactive waste shipments when they represent such a small percentage of shipments on the road today. Radioactive material incidents accounted for 15 of 6,176 reported incidents in 1988, or .2 percent, according to DOT. To answer public concern shippers and carriers must exercise extraordinary care in the conduct of their transportation activities and effectively communicate the degree of that care to the public.

Recommendations

In light of these concerns, NCSL believes that the following principles should underlie amendments to the Hazardous Materials Transportation Act:

1. The role of DOT should be more clearly defined. Specifically, DOT should be mandated to promulgate technical standards and criteria for the manufacture of hazardous materials containers, hazardous materials classification and definition, description, packaging, marking, labeling, documentation, and placarding of equipment for transportation. The requirements should be nationally uniform, and DOT should be directed to assume full enforcement responsibility.

2. DOT, in conjunction with the states, should be directed to promulgate performance and routing standards for hazardous materials shipments in commerce. This provision recognizes that unique local circumstances may justify the promulgation of certain state and local regulations. Such regulations should be allowed to stand if they met the performance standards and do not impose an undue burden on commerce.

3. There should be federal assurance of adequate and predictable financial assistance to support training of state enforcement personnel and development of adequate state data systems if the federal government shifts to the states responsibility for comprehensive enforcement of hazardous materials transportation regulations for shipments in commerce.

4. The federal government should assure that adequate and predictable funding is available for the development of state and local emergency response programs and for the training of emergency response program personnel.

5. State and local enforcement of the federal hazardous materials transportation regulations and state and local programs to deal with hazardous materials transportation incidents are dependent upon

information provided by shippers and carriers of hazardous materials. NCSL believes that the federal government should undertake a study to determine:

A. How such information is obtained (e.g., licensing, registration, notification, accident reporting requirements);

B. The relationship between state information requirements and states' ability to undertake enforcement and emergency response programs; and

C. The most appropriate level of government to collect such information.

Current proposals in Congress reflect many of these concerns.

States must also continue to take action to improve their capability in managing the transportation of hazardous materials. To assure a system which protects public health and safety, states that have not should adopt federal regulations governing the transportation of hazardous materials which include:

o   Uniform definition and identification of hazardous materials; and

o   Standards governing hazardous materials in commerce.

Additionally, states should establish:

o   Authority to adequately enforce the regulations;

o   A lead agency for hazardous materials enforcement and emergency response efforts; and

o   Authority to develop a comprehensive emergency response plan, encouraging private sector involvement.

Of course, NCSL understands the need for the free flow of commerce and the need for state and local governments to strive for greater regulatory uniformity with respect to hazardous materials transportation. However, such strides will only be made by state and local governments if they are assured that adequate and predictable financial resources will be made available to assure comprehensive enforcement of the regulations and the development of adequate emergency response programs.

Some positive developments in state/federal cooperation should be mentioned. The Commercial Vehicle Safety Alliance--a state alliance with support from the federal government--has had success in working for uniform regulation. Another initiative--the Cooperative Hazardous Materials Enforcement Development Program (COMED)--is also a promising development. The recently signed aid agreement between DOE's Idaho Operations Office and the State of Wyoming on cooperative radiological response activities is a further sign of progress. NCSL is encouraged by these developments.

To aid the ongoing state-federal dialogue in radioactive waste transportation, NCSL has formed a state task force on hazardous materials and high-level waste transportation. The group will meet for the first time in Washington, D.C., on May 19, 1990.

Proposed activities of the task force will include:

1. Monitoring the Congressional re-authorization of the Hazardous Materials Transportation Act.

2. Following transportation developments relative to the Department of Energy's monitored retrievable storage facility and the repository.

3. Working with the U.S. Department of Transportation, the Department of Energy and the other appropriate federal agencies to incorporate the state legislative perspective on issues related to emergency response to a transportation accident, planning for future shipments of high-level radioactive waste and other hazardous materials, and state involvement in inspection and permitting for trucks and trains carrying these substances.

References

National Conference of State Legislatures, State Issues 90: Priority Issues for State Legislatures, January 1990

_____, Hazardous Materials Transportation: A Legislator's Guide, February 1984.

Nuclear Waste Policy Act, as amended. 42 USC 10101, et seq.

U.S. Congress, Office of Technology Assessment, Transportation of Hazardous Materials: State and Local Activities (Washington, D.C.: U.S. Government Printing Office, 1986).

U.S. Congress, Senate Committee on Commerce, Science, and Transportation, Motor Carrier Safety and the Federal Highway Administration: Options Intended to Improve an Overloaded System (Washington, D.C.: U.S. Government Printing Office, April 1989).

U.S. Federal Emergency Management Agency, Guidance for Developing State and Local Radiological Emergency Response Plans and Preparedness for Transportation Accidents, August 1988.

DANGEROUS GOODS EMERGENCY RESPONSE
THE WESTERN AUSTRALIAN EXPERIENCE

K. Price *

Abstract:

A general overview of the State of Western Australia including:

- the legal framework of the Dangerous Goods and Emergency response management scenarios;

- particular problems unique to the Western Australian environment;

- what has been done to overcome those problems.

Part I

Western Australia has an area of about two and a half million square kilometres.

The demography of the State is such that the population is concentrated in the south-west corner of the State with isolated pockets, mainly associated with mineral development but also associated with agriculture, scattered throughout the State.

* Explosives & Dangerous Goods Division, Mines Dept., 100 Plain St., E. Perth, W. Australia 6004

This results in significant quantities of dangerous goods being moved around the State, generally over very long distances and through unpopulated areas. However, the historical development of the State is such that all roads went through the centre of all regional towns.

Typical dangerous goods which move through the State are dominated by fuel products including LP Gas.

However, complementary to the fuel products are:-

- Cyanide used extensively in extracting the 30 tonnes of gold produced each year;

- Chlorine produced at local plants;

- Ammonium nitrate manufactured in the State, the majority of which is used for on-site explosives manufacture;

- Pesticides used in agriculture; and

- Acids and alkalis used in gold processing.

Thus, there is a reasonable spread of dangerous goods, some in static storage tanks or manufacturing processes and others imported direct from offshore and transported through Western Australian ports to the point of use.

Legislation

Australia is a Federation of six states and several Territories.

The Federal Government is based in Canberra and is responsible for matters of national significance such as-

Defence
Communications
Customs
Immigration

Each State has its own Parliamentary system. Western Australia's is based in Perth and is responsible for

- (intra-state) transport
- education
- most infra-structure and resource management

Police and Fire Brigade are State Government responsibilities; thus there is a single police force responsible for law enforcement state wide. Similarly the WA Fire Brigade Board controls fire response state wide though volunteer brigades are used in many regional centres.

There is a local governmental system of limited power, which deals with local infra-structure, e.g.

    local roads

    sanitation

    recreation

    building standards

Naturally there is some overlap between these systems and they differ slightly from State to State.

The legislative framework for emergency response in Western Australia is State Government administered, based around the Explosives and Dangerous Goods Act which is administered by the Dangerous Goods Division in the Department of Mines.

This act gives legal authority to regulations to control the storage, transport and handling of dangerous goods and also the management of emergency responses.

Supplementary legislation includes

Statute	Major Thrust
Health Act	Waste Management
Environmental Protection Act	General environmental protection and restriction of major hazards generated from controlled
Occupational Health Safety & Welfare Act	Worker Safety
Sundry Emergency Services	Road Traffic Control, Fire Brigades, Ambulances, etc.

Regulatory Philosophy and Scope

The regulations contain provisions to facilitate emergency response and management of emergency scenes. These include provisions to:

- clearly identify vehicles (placarding);

- identify the nature and quantity of the cargo being carried (placarding and documentation);

- minimise the chance of small incidents escalating (first aid emergency response equipment and driver training);

∴ ensure 24-hour emergency advice is available for bulk transport;

∴ have public liability insurance

Parallel provisions are being built into the broader Dangerous Goods Regulations for all dangerous goods in storage.

Part II

Problems unique to Western Australia

The problems unique to Western Australia centre around the geography and demography.

The capital city of Western Australia is Perth with a population of around one million.

The capital is dominated by the Swan River and its nearby beaches. The surrounding hinterland is very pleasant as is most of the coastline along the State.

There are numerous small regional centres, a few with a population of around 20,000 but most with populations around 1000-5000. These tend to be associated with mining or farming operations.

There are long strips of road between these centres but generally there is only one route between them. Thus, there is not any real choice of routes if you wish to traverse the State.

Most dangerous goods are generated or imported through the south-west but are used in remote areas so transport is a major industry in the State.

When an accident occurs in remote areas, volunteers who rarely encounter an incident must attend to it. The incident tends to generate an enormous amount of local publicity when it occurs.

Action taken in the emergency response area in Western Australia is in three major thrusts.

1   Systematic response procedures.

2   Specific industry response.

3   Mutual aid groups based on commodity or geography.

Systematic Response

Around 1983 a system was developed to manage dangerous goods transport emergencies throughout the State. The Western Australian Transport Emergency Assistance Scheme (WATEAS) prescribed who would do what and when at a transport emergency involving dangerous goods.

Roles were defined for the ..

Organisation	To Do
Police	Crowd and incident
Fire Brigade	Control Combat
Mines and Health Depts	Technical advice
Carrier	Provide documentation, placards, driver training and hardware/materiel
Support Groups: (Water Authority, Electricity Commission)	As appropriate to their expertise

The WATEAS had a built-in provision for a review of every incident so that any problems could be monitored.

The system worked reasonably satisfactorily in the metropolitan area though there were a few problems in the country.

Typically it could take 6 - 12 hours for expert advice and suitable combat equipment to arrive at the scene.

Response in country areas was initially confined to recognising the problem containing it and awaiting further advice.

Following one particularly well-publicised accident involving the derailment of several fuel tanks, the State rail authority incorporated its rather ad-hoc emergency response procedures into the broader system and they now participate vigorously in the monitoring and development of that system.

A significant amount of sodium cyanide solution is carried by rail to railheads from which it is subsequently transported by road to various minesites. This is supplementary to the usual range of dangerous goods carried by rail. Consequently the development of more formal response procedures for rail is very appropriate.

Specific Industry Response

Isolated areas with a single dominant industry such as the North-West Shelf liquefied natural gas project were outside the emergency assistance scheme. This organisation had its own expertise which would probably be impeded by external emergency response groups. They developed a total hazard control plan which included a comprehensive emergency management procedures.

Similar systems are being developed for most other major industries in the State where dangerous goods are stored in sufficient quantity that an incident may extend beyond the boundaries of the plant. However these continue to depend heavily on the local fire service.

Total Hazard Control Plans

It is fair to say that for major plants, emergency response planning in WA now begins at the conceptual design stage.

We expect that new plants should be designed with safety at the forefront and with due adherence to hazard and operability procedures at all stages of development.

Inspectors of the Department of Mines participate in Hazop studies and when plants are constructed they must have in place a Total Hazard Control Plan which takes into account such things as:

- failure scenarios
- alarm procedures
- evacuation and response procedures and
- emergency shutdowns

Mutual Aid

A mutual aid group has been developed for vehicles transporting solid cyanide around the State. With a gold industry which produces about 30 tonnes of gold annually, some 60 000 tonnes of solid cyanide are carried from the port and rail head near Perth out to the Goldfields over distances for 250 to 1000 km. The mines which use the cyanide have expertise to deal with any uncontrolled release and have the hardware to effectively manage an emergency.

By identifying all significant mines which have processing plants and having them nominate a responsible officer, there existed the basis for a mutual aid group. This information is coordinated by the Western Australian Fire Brigade who arrange for local expertise to deal with any cyanide spillages in or near the goldfields.

The initial reluctance to participate in mutual aid groups of this kind (because of liability fears) hopefully will be overcome by the introduction this session of Parliament of a "Good Samaritan" clause into the Explosives and Dangerous Goods Act.

A more significant mutual aid system is being developed for the Kwinana area south of Perth. This area contains some 2/3 of the State's major chemical based industry and the potential savings from sharing emergency response equipment are significant.

Effects of the Emergency Response System

The Division became more actively involved in far more accidents when the legal responsibility to report came into effect.

This resulted in the generation of more reliable accident statistics which became a database to justify changes in the regulations.

For example, at an early stage it pinpointed that far too many accidents were being caused by packages of dangerous goods becoming dislodged from trucks. This resulted in an amendment to require effective restraining devices to be fitted to all vehicles carrying dangerous goods.

Indeed, the reporting system itself became a valuable resource. In an ideal world, there would be a common accident data base and the statistics generated would be used to assist decision making in a consistent manner in all nation states.

In practice, there is not yet a uniform reporting system and data base across all Australian states. Western Australia has the only effective operating system in Australia - loosely based on the US system but with a data base too small for national purposes. It is gaining some acceptance in other States.

Outstanding Problems

There is no way that it could be said that all the problems are resolved at this stage. Yet to be resolved are:-

1   Compensation - who meets the costs of a cleanup. Philosophically, action at a scene can be divided into response and recovery phases. (Emergency) response ends when the area is declared safe but recovery (usually environmental cleanup) may take much longer.

  As the entire community pays the emergency services to be there, claiming costs from carriers following an accident may not be morally justifiable. However recovery costs should be met by the perpetrator and the mandatory public liability insurance carried by all transporters should be used for this. No response or recovery group has yet made a claim for compensation for expenses incurred in a dangerous goods emergency in Western Australia.

2 Protection of underground water resources which supply significant quantities of the water used in the Perth metropolitan region.

Australia is the driest continent and underground water resources are critical to our long-term survival. Action is being taken to improve long-term cleanup procedures.

3 Responsibility for advice for small incidents.

A typical example would be a nationwide retail company which gets a consignment including, for example, pool chemicals, cleaning solvents, food, gardening pesticide and a large quantity of general merchandise. This may be despatched from the supplier through a transport yard and all packed on one vehicle which, if it is involved in a fire, creates problems for the emergency services.

Where should they go for advice?

Neither the carrier nor the consignor will have the necessary expertise. The supplier is likely to be 5000 km away. Typically government advisers attend and sort out the problem.

This is satisfactory only if they can be recompensed.

Route planning. This is compounded by the problem of the single route available for most long distance movements and also by the usual problem of local deliveries. In my opinion, no effective dangerous goods vehicle routing systems exist though some are better than others. Steps are being taken to amend the Western Australian legislation to make our vehicle routing controls as effective as the best in the world.

Conclusions

At the commencement of the eighties Western Australia had:

- no real system for dealing with dangerous goods;
- had no legal authority for dealing with emergencies; and
- few dangerous goods other than flammable liquids.

At the commencement of the nineties we have:

- a vast range of dangerous goods being carried;
- a good system to deal with emergencies though some gaps still remain;
- operating procedures backed by a sound legal framework; and
- a first class regulatory system which includes sound inspectoral procedures based on a philosophy of advice and accident prevention rather than enforcement and punishment.

RISK PERCEPTION

*Risk Management in the Transportation of Dangerous Goods
--the Influence of Public Perception "A Discussion"*

Marjorie K. Matthews[1]

Abstract

In 1988, a fire at Saint-Basile-le-Grand, Quebec involving a PCB (polychlorinated biphenyls) warehouse sparked continuing controversy which included a two-week evacuation of residents. When attempts to remove the PCB waste from storage for ultimate disposal, public reaction and the NIMBY (not in my backyard) syndrome prevailed and interrupted the handling and transportation of this which, from a risk management viewpoint, would have provided minimal social impact.

This is just one of many examples whereby public perception of risk can unduly influence the proper response to effective risk management, in this case, for the transportation of dangerous goods. However, risk managers also have an equal responsibility to ensure the adaptation of public or social values with respect to responsive and credible decision-making. The management of risk on strictly addressing probability and consequences, without taking into account public concerns or anxiety, will invariably result in a foundation which may be scientifically correct, but lacks social fibre. As a public service, the two are inconsistent.

Introduction

As a government regulator for the transportation of dangerous goods, Transport Canada has the responsibility to ensure that these products are transported safely for the social and economic good of the country. As a result of the fire involving PCB´s at St-Basile-le-Grand, government authorities initiated efforts to arrange for the removal and ultimate disposal of PCB´s in storage. At that time, there were limited facilities for portable incinerators; therefore, the product had to be transported from Quebec to an approved disposal site--either in Alberta or in Europe.

---
[1]Director, Risk Management Branch, Transport Dangerous Goods, Transport Canada, Ottawa, Canada, K1A 0N5

Discussion

Due to the controversy and negative publicity that almost immediately befell government when they attempted to organize proper methods of containment and transport, the NIMBY (Not In My Backyard) syndrome essentially prevented a hassle-free operation. Faced with an unbelieving public, governments brought in independent authorities to assure credibility and soundness of the procedures. "Expert" opinion on the subject varied from one extreme to the other, further compounding the public's mistrust in the handling of this crisis.

This leads into the topic of my paper--that of the influence of public perception in the management of risk. There are many formulas to describe the concept of risk. The most common is the probability of an unwanted event weighted by the consequences of that event. For dangerous goods, the unwanted event is the accidental release of the product; the consequences of an uncontrolled release (both potential and real) are the effects of the nature of the danger or hazard posed (i.e. flammability, toxicity) in terms of damages and losses incurred. During transport the risk may be even more pronounced due to the uncertainty of not only when, but where an accidental release of product may occur. The only assurance of achieving zero risk for the transport of dangerous goods is to reduce one of the two components, probability or consequence, to zero. Reducing the probability of a transportation accident to zero requires prohibition of the movement. Dangerous goods are transported in response to economic demands which are generated by society. We, therefore, have completed the circle in attempts to manage risk for an activity generated by society and are now introducing a public responsibility in this process.

Government is expected to achieve a balance between the safe transport of dangerous goods at a reasonable cost. To ensure the levels of safety that the public **appear** to demand would effectively constrict an affordable operation and may not represent a social benefit in terms of a net contribution to the economy. It then becomes essential to include public perceptions and opinions as a monitor on the adequacy of regulatory control for transport of these products. Once the risk manager has arranged for the incorporation of these views, it becomes a continuing responsibility to ensure comprehension of risk values and assessments and the confirmation of public priorities. This is accomplished through risk communication.

Communication of risk should not concentrate on the presentation of probabilities since, unless it is related to an activity that the public can understand, it is meaningless. Risk comparisons are only sound and convey the proper impact when the examples used are common in terms of voluntary vs. involuntary choice; rare vs. frequent events or when the potential or expected consequences are similar with respect to magnitude (catastrophes vs. singular death or injury).

In responding to perceptions of risk, risk communication should focus on the potential for disaster, the treatment of uncertainty, the methods of societal control and the distribution of benefits. It must be understood though that not all of society has the same values and concerns with respect to acceptance of risk. This then provides a challenge to the risk communicator to address the range of boundaries which represent tolerant societal values. It would be naive to expect that an acceptable response could be generated to all.

It is also necessary to establish a process whereby the risk communication invokes a two-way exchange of ideas and information. Communicating such to the public is intended to solicit their reactions and priorities. Regulators may then address public opinion through the alteration or modification of the final analyses, if the views so gleaned provide further insight to the issue at hand. Arguments must be clear and respond to the concern. Conflicts in the message will only serve to confuse if not anger the audience. A balance must also be provided to demonstrate opposing, yet equally valid opinions and also serves to promote credibility in the message.

Conclusion

As a risk manager, it is not enough to merely deal with the quantitative aspects of the risk equation. The interpretation of the numbers so calculated must take into acount the effects of qualitative values and opinions expressed by the public and often monitored by the media. This two-way process of risk communication demonstrates to the public not only the importance of their input but also serves to establish a framework whereby the decisions made are truly reflective of the priorities and values indicated by Canadian society; and allows the public to influence, if not have a role, in the overall management of risk in the transportation of dangerous goods.

It is not the intention of this paper to state that the inclusion of and reaction to public attitude always improves the outcome of risk management decisions because of the risk transfer phenomena; however, government regulators have a way to go to educate and inform the public that they have a right, as well as a responsibility, in the development of decisions for the management of risk in the transportation of hazardous materials.

References

Canadian Emergency Preparedness College, *The Social Costs of Disasters*, Symposium '89, Arnprior, Ontario, September 1989.

Institute for Risk Research, *Issues and Challenges in Risk Communication and the Mass Media*, IRR, Waterloo, Ontario, March 1990.

National Research Council, *Improving Risk Communication*, National Academy Press, Washington, D.C., 1989.

The Toronto Area Rail Transportation of Dangerous Goods Task Force, Final Report, Minister of Supply and Services Canada, Ottawa, Ontario, 1988.

DEVELOPING HIGH-RISK SCENARIOS AND COUNTERMEASURE IDEAS
FOR MITIGATION OF HAZARDOUS MATERIALS INCIDENTS
by
Eugene R. Russell, Sr.[1]

Abstract

Kansas State University (KSU) conducted a comprehensive study of the development of a set of prioritized, extreme-risk scenarios, the development of a set of feasible, practical and implementable protective systems, and a report to summarize guidelines on the use of these protective systems to mitigate potential, extreme-risk situations that could occur during the transport of hazardous materials (hazmat) on our highway system. This paper covers the methodology used to complete the first two objectives with use of a state's panel.

The research study was limited to materials spilled within the highway system. It focused on potential risks which would result in severe, long-term, permanent, irreparable or catastrophic consequences, and existing technology and state-of-the-art knowledge for development of protective systems to mitigate these consequences. The protective systems within the scope of this study were systems constructed or physically incorporated into the highway system or modifications thereto.

Introduction

General. Early in the project it was concluded that designing protective systems into highway systems specifically to prevent or mitigate hazardous materials spills was a new and unique concept. No U.S. literature existed directly addressing this concept. Literature on systems that could be adapted to the concept, such as

---
[1]Professor of Civil Engineering, Kansas State University, Manhattan, KS, Member ASCE.

drainage containment systems, high-strength barrier rail, etc., was available, but first the catastrophic nature of spills had to be defined.

By its nature, a catastrophic occurrence resulting from the highway transportation of hazmat is a rare event. Not everyone agrees to a universal definition of "catastrophic occurrence." Catastrophic can mean many things to different people. Developing catastrophic scenarios meaningful to the states, leading to concommitant, feasible, practical protective systems useful to the states and credible with states' personnel, was an early stumbling block. All sorts of catastrophic scenarios could be dreamed up, but would they be realistic or meaningful?

It was decided to contact all states through official Federal Highway Administration (FHWA) channels and form a project advisory panel of personnel from those states wishing to participate in the research. This panel was then used to develop potentially catastrophic scenarios. This approach assured the scenarios developed were real state concerns and not merely creations of overactive imaginations of the researchers. Eleven ranked catastrophic scenarios were developed.

The next task was to develop protective systems keyed to the scenarios. These protective systems were to be feasible, practical, and implementable. In this age of modern technology, almost any system imaginable is feasible and implementable -- given unlimited resources. Obviously states do not have unlimited resources; therefore, what is "practical" will ultimately be a state policy decision based on some sort of cost-effectiveness analysis. Once again, to come up with a range or a set of possible, credible protective systems, the states' panel was used. This ensured that the set should not be too far outside the range of what an individual state decision-maker would consider practical or at least worth investigation.

Scenario Development

Using a states' panel assured the scenarios developed would represent real concerns to the states. Biases would exist due to the varied background of the individual panel members, their experience, and the varied experience of the individual states in regard to hazmat flows. However, because the research was to address a cross-section of states' problems, <u>as the states perceived the problem</u>, it was appropriate to develop the scenarios with whatever inherent biases exist in the real world of varied state concerns. No two states are likely to have the same concerns, and no two states would be likely to rank a set of catastrophic scenarios exactly the same, nor even all agree on the same set. The fact that a definite consensus was arrived at was, in itself, a major accomplishment.

The advisory panel was asked for real or hypothetical scenarios of concern to their state. Approximately 60 separate, independent scenarios were suggested. Through several rounds of rankings, the original 60 were narrowed down. Most fell into a few categories such that by generalizing a statement covering the category, several of the original scenarios fit each statement. Thus, the process of narrowing the scenarios down was a process of fitting all 60 into 11 general scenario statements, not picking the top 11 and discarding 49.

For purposes of the rating process a "Key to Scale Values," shown in Table 1, was developed to aid the responders.

The final round asked for "approval" of the list of 11 secnarios on the following page, Table 2. Although some minor changes were suggested, 28 of 30 respondents approved the list as presented. All but one or two praised the resulting set of scenarios and the conclusions. Thus, the authors are certain the list reflects a valid set of real state concerns.

The ranked set of 11 should not be considered to have a scale -- neither a relative nor an absolute scale. Individual states could develop state-specific criteria to do a meaningful ranking of this list or a similar state-specific list.

One part of the study tied highway geometric components to the potentially catastrophic situations. Table 3 shows an example of the results. The greatest perceived danger is from elevated facilities, followed by depressed facilities with development over (air-rights structures) and, lastly, receptors* adjacent to the facility. A sample of the results are shown in Table 3 with a summary shown in Table 4.

---

*Here, receptor is defined as the population, property, etc., subjected to a hazmat release consequence. As a mater of interest, these adjacent facilities are rated in order of their individual scores as rated by the panel.

TABLE 1. KEY TO SCALE VALUES.

Scale Value*	Key
1	**Very minor incident**; of little or no consequence under normal conditions.
2	**Minor incident**; little chance of escalation, little danger to life or serious or long-term environmental damage (aquifer, reservoir, or water supply) unless grossly mismanaged.
3	**Potentially dangerous incident**; but not likely catastrophic, danger to life or environment (aquifer, reservoir, or water supply) only if not handled properly.
4	**Neutral**; no clear catastrophic potential yet hard to predict.
5	**Definitely dangerous incident**; could be catastrophic under certain conditions of traffic, weather, or inadequate response. Could easily escalate to catastrophic situation.
6	**Very dangerous incident**; high catastrophic potential, high probability of loss of life, serious injury, or long-term damage to environment (particularly aquifer, reservoir, or water supply).
7	**Definitely catastrophic incident**; loss of life, serious injury, serious damage to environment (particularly aquifer reservoir, or water supply) is certain to be avoidable only with extreme good luck.

*In general terms, where all replies are averaged, a mean value greater than 4 was interpreted to mean the scenario is catastrophic or has catastrophic potential.

Developing Protective Systems Keyed to the Scenarios

Philosophy. After ranking the 11 prioritized extreme risk scenarios, the next task was to develop "feasible, implementable, and practical" protective systems keyed to the 11 scenarios.

TABLE 2. RANKED, GENERALIZED SCENARIOS.

Rank	General Scenario Description
1	Poisonous, toxic, flammable or explosive material endangers large numbers of trapped motorists, e.g., between interchanges, in cut section, or in traffic jam downwind of poisonous or toxic gas release.
2	Chemical spills of poisonous or explosive materials that could enter underground "METRO" stations or transit tunnels through sidewalk vents, etc. (Includes entry of lighter-than-air toxic or poisonous gases into adjacent or overhead transit stations.)
3	Hazardous materials accidents causing release of toxic, flammable, or explosive materials in tunnels.
4	Gasoline, LNG, propane (flammables, explosive gases), etc., accidents and releases on elevated facilities, including ramps thereto, with people at risk below or in adjacent buildings.
5	Release of poisonous, toxic, or explosive gases in populated areas in general and/or in locations and situations where special populations and/or institutions such as schools, hospitals, hotels, nursing homes, apartment complexes, etc., are at risk.
6	Release from accidents between hazardous materials containers on highways and passenger trains or trains carrying hazardous cargo either at rail-highway crossings at grade or in situations with shared rights-of-way, such as freeways with transit in the median.
7	Explosive materials in facilities in populated areas, and particularly in situations and areas where catastrophic consequences could occur to highway structures or apartments--adjacent or on air rights. Includes situation with adjacent petro-chemical plant that could result in conflagration.

HIGH-RISK SCENARIOS

TABLE 2. RANKED, GENERALIZED SCENARIOS (continued).

Rank	General Scenario Description
8	Sufficient quantities of poisonous materials, such as herbicides or dangerous biological/agents (or any material causing long-term or permanent damage) being released into a potable water supply, particularly reservoirs and susceptible aquifers and/or watersheds.
9	Rural, hilly, or mountainous areas with cities or towns at bottom of long or steep grades where brake failure of hazardous materials carriers could cause catastrophic consequences to the populated area.
10	Spills of nuclear wastes or other nuclear materials, particularly in populated areas, areas affecting water supply, or areas particularly difficult to respond to and/or clean up.
11	Carriers of toxic, flammable, or explosive materials leaking material during transit in heavily populated or congested areas.

The original protective system ideas came from the states' advisory panel. The panel suggested and evaluated 98 original protective system ideas.

Editing was very slight so as to keep the ideas essentially as the panelists had presented them. Several had to do with regulatory-type solutions, which were outside the scope of the project. However, the panelists had been informed of this many times, and the fact that many still felt these sorts of solutions were best, in the opinion of the author, is significant information.

A 1 to 7 rating scale was used as shown in Table 5.

The responses were so varied that it was not immediately clear how to interpret the results. For example, the results of the mean response for all protective systems asociated with scenario No. 4 varied from less than 2.0 to 5.3. Large differences appeared between responders to the same protective system for the same scenarios. The range on almost all individual

TABLE 3. PARTIAL RESULTS OF THE RATING VALUE FOR CATASTROPHIC POTENTIAL OF VARIOUS HIGHWAY SEGMENTS FOR GASOLINE.

Gasoline: faculty descriptor/reactor catastrophic potential response

<u>Rating Summary*</u>　　<u>Urban Freeway Components</u>
Mean

a. elevated

	(1)	basic segment
5.52		(a) over shopping center
5.04		(b) over parking
3.93		(c) no development under
	(2)	weaving area (non-ramp)
5.52		(a) over shopping center
5.04		(b) over parking
3.93		(c) no development under
	(3)	ramp/ramp junction/acel.-decel. lanes
5.52		(a) over shopping center
5.04		(b) over parking
3.93		(c) no development under
5.08	(4)	drainage (from el.) to storm sewer
5.04	(5)	drainage (from el.) to combined sewer

b. at-grade

4.44	(1)	basic segment
4.44	(2)	weaving area (non-ramp)
4.33	(3)	ramp/ramp junction/accel.-decel. lanes
4.83	(4)	drainage into storm sewer
4.93	(5)	drainage into combined sewer

c. depressed

	(1)	basic segment
5.15		(a) with air-rights development
4.37		(b) without air-rights development
	(2)	weaving area
5.22		(a) with air-right development
4.44		(b) without air-right development
	(3)	ramp/ramp junction
5.19		(a) with air-rights development
4.41		(b) without air-rights development
4.67	(4)	drainage into storm sewer
4.70	(5)	drainage into combined sewer

TABLE 4. GENERALIZED SUMMARY OF ROUND 4 RESULTS.

Rank	Approx. Avg.* Mean Score	Generalized Highway Facility
1	5.6+	Elevated facilities over development
2	5.5	Depressed facilities under development
3	5.0 to 5.4	Any facility adjacent to vulnerable population in order of: a) nursing home or hospital b) schools c) apartments d) shopping center e) hotel f) factory g) hazmat storage facilities
4	4.0	Drainage into sewage system

*Based on 1-7 scale explained in Table 1.

TABLE 5. KEY TO SCALE VALUES.

Scale Value*		Key Guidelines to Assist Raters
Bad (Worst)	1	Nearly impossible to implement, not at all practical, will serve no useful purpose
	2	Very difficult to implement; little value
	3	Difficult to implement; some value possible but probably not worth the effort or cost
"Neutral"	4	Hard to judge; not clearly "good" or "bad" idea
	5	Possible merit as practical and implementable protective system; worth further thought or development
	6	Clear cut merit as practical and implementable protective system
Excellent (Best)	7	Highly feasible, very practical, useful and efficient, i.e., excellent and very desirable

*In general terms, where all replies are averaged, a value less than 4 would suggest that the idea would be highly difficult to design/construct and install or would not be very useful/desirable; i.e., throw it out.

protective systems was 1 to 7. A similar pattern could be found in the mean responses to all protective systems.

Individual responders were studied and the only trend noted from this exercise was that the highway engineering-types generally rated higher than the group mean and the environmental/responder types generally rated below the group mean.

It clearly came down to using a mean that reflected the various biases of the responders vs. using any one individuals' or groups' expert opinion or weighted opinions. It was concluded that these fluctuations represented real-world differences of opinions in a new area. It was decided to rate the protective systems using the collective panel mean, with all its inherent biases.

There was no rational way to determine anything but an arbitrary "high/low" cut-off point to reduce the number of protective systems to a set that could be handled well with the available project resources. A mean rating of 4.0 was chosen as the cut-off point.

The results for Scenarios 1 and 4 are shown in Table 6. Communication and detection systems ranked highly for most of the scenarios and these were ranked separately. These results are summarized in Table 7.

Results. One general conclusion could be readily seen from the results. Communication and detection type protective systems were rated very high. Communication and detection systems would be site-specific protective systems, i.e., high-risk, high catastrophic potential situations of a particular nature and not for general use. Much research and development is needed in this area if these systems are to be adapted to general highway use.

Regulatory-type Responses. Even though it was made clear to the panel that regulatory-type solutions to the scenarios were not within the scope of the project, almost as many were returned and ranked generally higher than protective system-type ideas.

The overall mean rating of the regulatory-type was 5.3, compared to 4.7 for the protective system type. Response to nine of the individual scenarios show that the regulatory-type were rated higher in every case.

HIGH-RISK SCENARIOS

TABLE 6. PROTECTIVE SYSTEM RATING RESULTS ALL PROPOSED PROTECTIVE SYSTEMS RATED 4.0 OR GREATER FOR SCENARIO 1 AND SCENARIO 4.

SCENARIO 1 -- Poisonous, toxic flammable or explosive material endangers large numbers of trapped motorists, e.g., between interchanges, in cut section or in traffic jam downwind in poisonous or toxic gas release.

Protective Systems-type Solutions
Rank	Mean	
1	5.1	Traversable medians
2	5.0	Emergency phone call boxes on all hazardous cargo routes
3	4.7	Crossovers
3	4.7	Median openings
4	4.6	Highway exits designed for traffic entrance (response team) from opposite direction
$\bar{X}$ = 4.8		

Regulatory-type Solutions
R1	5.2	Routing restrictions
R2	5.0	Prohibition on hours (curfews)
R2	5.0	Prohibit large trucks through congested areas (routing)
$\bar{R}$ = 5.1		

SCENARIO 4 -- Gasoline, LNG, propane (flammables, explosive gases), etc., accidents and releases on elevated facilities, including ramps there-to, with people at risk below or in adjacent buildings.

Protective Systems-type Solutions
1	4.9	High performance barrier/rail systems, to prevent such an accident
2	4.8	Avoid use of open rails on structure
3	4.4	Robust drainage with holding reservoirs that can be isolated from regular storm drains (and later pumped) should a spill occur
4	4.2	Conduit railing for automatic spraying of water
4	4.2	Relocate or close ramps in critical locations; install improved barriers; prohibit truck use of ramps
$\bar{X}$ = 4.5		

Regulatory-type Solutions
R12	5.3	Reduced speed
R2	4.3	No hazmat through urban area (prohibition)
$\bar{R}$ = 4.8		

TABLE 7. SUMMARY OF TOP FOUR COMMUNICATION AND DETECTION SYSTEMS FOR ALL SCENARIOS.

A. Detection

1. Non-Remote Sensing Techniques, general:

Mean	Rank	Specific Examples
5.9	1	Explosimeters
5.8	2	Colorimetric Indicators
5.4	3	TLV Sniffers
5.1	4	Water Analysis Kits

2. Remote Sensing, general:

Mean	Rank	Specific Examples
6.7	1	Gaseous-measuring laser radar systems
6.5	2	Plume and Haze analyzer
6.5	2	U.S. Army's remote sensing XM 21
6.0	4	Correlation spectrometer

B. Communication

Mean	Rank	Specific Examples
6.3	1	Instructions with all drivers
5.5	2	Radiation indicators on trucks
5.4	3	Instructions pasted on truck
5.3	4	Posted standard instructions regarding nature of hazard, preliminary protective measures, and first aid

Summary and Conclusion Re Protective Systems

Based on responses of a large panel representing a broad cross-section of states' concerns, regulatory-type preventative measures dominated suggested solutions. Conversely, it can be concluded that the physical, protective system concept is not applicable as a general preventative or mitigating approach. It is limited to a few site-specific, high-risk situations where the protective system approach is clearly effective and the risk is deemed high enough to offset the cost. This is a policy decision of each individual state and is the heart of the "practicality" criteria.

The study came up with only one type of protective system that could be called "preventive." This type consists of various barriers to contain a hazmat vehicle on or within the roadway to prevent it going off an overhead facility, off a ramp, into a school yard, etc.

HIGH-RISK SCENARIOS

Various types of barrier rail designed to contain large trucks would be typical of this category.

All others can be classified as "mitigating." This type dominates the responses, including categories such as detection and warning systems, systems to facilitate escape and response, systems to mitigate fire/explosion consequences, systems to mitigate spill consequences and systems related to highly specialized situations, e.g., elevated METRO vents. Protective systems should fit predominately into one category. They are so categorized in Table 8.

Finally, considering all input into the scenario development and ranking protective system survey and rating, it was concluded that the key to the guidelines is a report pointing out areas of general systems that should be considered, rather than a design manual that attempts to set forth specific design and/or standards that must be followed.

Currently, only a few protective systems appear to have clear cut merit in reducing risk of hazmat spills and/or mitigating consequences. The most promising, and most applicable, to many high-risk situations are in two general categories: 1) high-strength rail to keep hazmat vehicles within the highway system in the event of an accident and 2) closed drainage systems to contain or control spilled material that may result from an accident. These and others should not be overlooked, as preventing one catastrophic occurrence where hundreds of lives could be lost is worth considerable effort. The primary goal of the guidelines manual is to ensure that high-risk situations with catastrophic potential are not overlooked.

A brief summary of protective systems are discussed in detail in a guidelines report. The guide presents a suggested risk model states can use to define or flag potentially high-risk situations and to identify data needs to evaluate the alternatives to determine if a protective system should be designed and implemented. Appropriate types of protective systems to reduce that risk are presented for various generalized scenarios.

The guidelines report will give the state designer/planner or administrator enough information to alert him/her to situations where there is high risk of a catastrophic occurrence and to assist him/her in the decision-making process of determining whether a protective system would be appropriate.

TABLE 8. CATEGORIZATION OF PROPOSED PHYSICAL, PROTECTIVE SYSTEMS FOR HIGHWAYS

Category	System
	I. MITIGATING
A. Detection and Warning	Built-in PA systems Emergency call boxes Gas detectors/alarms Monitoring for quick response Communication/detection systems
B. Systems to Facilities Escape and Response	Crossovers Transversable medians Highway exit/entrance redesign for emergency response vehicles Emergency exits with heavy doors (tunnels) Arrows pointing to nearest exit (tunnels)
C. System to Mitigate Consequences	Foam blanketing systems Fire/Explosion Large sprinkler systems Effective vent systems
D. Systems to Mitigate Spill Consequences	Pea-style vents to trap gases Effective vent systems (closed areas) Robust drainage with holding reservoirs Avoid use of open rails on structures Large sumps Grease trap sedimentation basins Floating surface barriers Drainage gutters directed toward collection points Retention basins that automatically close Clay blankets or barrier membranes
E. Specialized Situations	Fresh air vents at elevated levels (METRO) Coamings over street-level intake vents (METRO) Air intake away from roads (tunnels, METRO) Massive barriers with energy absorbing material (runaway trucks

TABLE 8. CATEGORIZATION OF PROPOSED PHYSICAL, PROTECTIVE SYSTEMS FOR HIGHWAYS (continued)

                    II. PREVENTATIVE

A. Containment   High performance barrier systems
B. Control   Truck escape ramps
                        Upgrade truck runoffs
                        Wide shoulders

Acknowledgements

The work reported in this paper was conducted under sponsorship of the Federal Highway Administration. However, the findings and conclusions in this paper are those of the author and do not necessarily represent the views of the Federal Highway Administration.

References

1. Russell, Eugene R., Sr. "Protective Systems for Spills of Hazardous Materials, Volume I: Final Report," Federal Highway Administration, FHWA-RD-89-173, May 1990.

2. Russell, Eugene R., Sr. "Protective Systems for Spills of Hazardous Materials, Volume II: Guidelines," Federal Highway Administration, FHWA-RD-89-174, May 1990.

3. Harwood, D. W., and E. R. Russell, "Present Practices of Highway Transportation of Hazardous Materials," Federal Highway Administration, FHWA-RD-89-013, November 1989.

4. Harwood, D. W., E. R. Russell and J. G. Viner "Characteristics of Accidents and Incidents in Highway Transportation of Hazardous Materials," presented at the 68th Annual Meeting of the Transportation Research Board, January 1989.

Risk Perception and Policy Preferences of Fire Chiefs:
A Survey of Hazardous Materials Issues

Seymour Warkov[1]

and

Kostas G. Zografos[2], A.M., ASCE

Abstract

This paper presents the results of a survey of Connecticut fire chiefs and discusses how their perceptions of risk affect their attitudes towards risk preparedness and risk management practices. The results indicate that support for aggressive intervention on the part of local governments in the hazmat transport system is a function of risk perception, population size and low ratings of system effectiveness.

Introduction

This paper reports the results of a survey of Connecticut fire chiefs and their views concerning hazmat transport system issues. We will describe some findings concerning fire chiefs' perceptions of hazardous materials; and examine the relationship between risk perception and selected policy issues; and note variations in response on the basis of community size and types of highway system.

Data pertaining to these themes are drawn from a state-wide survey of a randomly selected sample of 95 fire chiefs in the State of Connecticut who were interviewed by telephone during Spring, 1989 (Warkov, 1989). The telephone interview schedule measured fire chiefs' perception and experience with various aspects of hazmat

---

[1]Professor, Department of Sociology, University of Connecticut, Storrs CT 06268

[2]Assistant Professor, Department of Civil and Architectural Engineering, University of Miami, Coral Gables, FL 33124

transport in the State, including hazards assessment activity, restricted routing and other issues.

Based on a list provided by the State Fire Administrator, interviews were completed with chiefs of fire departments serving the State's 21 largest cities and towns, while another 74 were conducted with a 50% sample of fire chiefs selected from the State's other 148 towns. This stratified random sample can be "weighted" up to represent all 169 towns in the State (See Warkov, 1989).

Research Rationale

Concern about the safety of the public have prompted a number of investigations of the hazmat transport system. This project focuses on the concerns of one set of key actors in the system, namely chiefs of local fire departments, officials who assume command functions on arrival at the scene of a hazmat accident. Chiefs interviewed by the project staff direct fire departments in some 60 small towns (1980 pop. up to 7,500); 89 mid-sized towns (7,500-39,999) and 20 "big cities (40,000 and over).[1] Based on the State of Connecticut Functional Classification System, two thirds of the sampled towns are located in the path of an expressway, and 67% have at least one "principal arterial" highway. In combination, nearly half (46%) of the towns have both an expressway and a principal arterial highway, while another 40% have one or the other.

This paper explores the relationship between perception of risks associated with the transport of three hazardous materials (gasoline, propane and sulfuric acid); risk preparedness; risk management, especially restricted routing, and perceived effectiveness of the hazmat transport system. For example, we would expect that the assignment of low effectiveness ratings to certain features of the hazmat transport system would foster a preference for more aggressive intervention on the part of local governments than would be the case if fire chiefs were relatively satisfied with current system performance.

Risk Perception

Risk perception measures were derived from scenarios varying the material to be assessed for risk level.[2]

---

[1] Also see: Zografos and Warkov, 1989.

[2] The section describing measures of risk perception, risk management, and restricted routing is excerpted from Warkov, 1989:11-13.

I'm going to ask you to think about some kinds of accidents that might occur with hazardous materials to be assessed for risk level.

Suppose a cargo tank truck transporting gasoline tipped over and ignited, and spilled its load on a designated route in your town.

How much risk would the gasoline spilled pose to the first responders? A high risk, a moderate risk or a low risk?

How much risk would the gasoline spilled pose to nearby residents? A high risk, a moderate risk or a low risk?

How much risk would the gasoline spilled pose to the environment? A high risk, a moderate risk or a low risk?

Some 70% of the fire chiefs rate this scenario as entailing "high risk" to first responders; 63% also rate the risk as high for nearby residents; and 66% consider the gasoline a high risk to the natural environment. The second scenario focused on corrosives and was framed in the following terms:

Now suppose a cargo tank truck transporting sulfuric acid tipped over and spilled its load on a designated route in your town?

Fire Chiefs almost unanimously (96%) rate this scenario as entailing a high risk to first responders, to nearby residents (91%) and to the environment (89%). The third scenario concerned propane, a fuel commonly used to meet the energy requirements of rural households.

Now suppose a cargo tank truck transporting propane tipped over and spilled its load on a designated route in your town.

In this case, fully 88% fire chiefs perceived this fuel to constitute a high risk to first responders; nearly as many (86%) said the same for nearby residents. However, substantially <u>fewer</u> (53%) deem spilled propane to entail a high risk to the environment.

Risk Preparedness

Effective responses to hazmat transport accidents/incidents presumes access to information that facilitates accurate risk assessments. According to our sample of fire chiefs, 79% of the towns have asked local businesses and industry for assistance in gathering this information and 51% of the towns have used part-time workers and/or volunteers in this capacity. Another 59% have obtained information on routes used to transport hazardous materials; 52% know about the types of hazardous materials being transported in or through the town; 37% have information concerning the risks from moving different sorts of hazardous materials; 36% have data on container types used; 32% say the town has collected information on the time of day such materials are transported, and 28% know something about the volume of hazardous materials being transport in or through the town. Finally, 21% indicated that their town has gathered information on how often such trips are made. In sum, Connecticut towns are most likely to have information on <u>routes</u> used to transport hazmat; even so, 31% have not yet compiled these critical data. To be sure, cities and towns vary in their capabilities to gather and translate such information into viable programs. Nevertheless, these findings point up a deficiency in the hazmat transport system.

Restricted Routing-An Example of Risk Management

Emergency planning and preparedness increasingly rely on other risk management practices such as the placing of restrictions on the routing of the hazmat transport. Fire chiefs are one set of actors whose judgements are essential for setting restricted routing practices.

Nearly all (88%) support the idea of having a town ban hazmat transport during peak periods of traffic, and with one exception, very strong majorities favor the remaining restrictions as well. For example, 78% would have the town restrict vehicles transporting hazardous materials to specific routes; 76% would require permits for carriers of hazmat; 75% would require permits for shippers of hazmat; 70% would require a special lower speed limit for vehicles carrying hazmat; and 68% would ban the use of roads or bridges for specific materials. However, only 36% would provide escorts on routes used for moving hazmat through their town, an option that a number volunteered was not economically viable for their towns. Since fire chiefs favor safety over prices by a margin of 93-5 when asked to choose between the two, these findings suggest that safety is compromised in the face of local fiscal pressures. On the other hand, very strong majorities (between 67-78%) favor prohibiting such transport during heavy traffic periods, requiring permits for shippers and carriers of hazardous materials, requiring a special lower speed limit for vehicles carrying hazardous materials, and banning the use of roads or bridges for specific materials.

Provision of escorts would entail direct costs and fire chiefs are reluctant to impose such costs on their employers. Consequently, we found, as noted earlier, that only 36% favored this form of restricted routing. Does risk perception influence their position on this question? As shown in Table 1, it does: what fire chiefs are responding to is <u>perceived risk to nearby residents</u> and this seems to operate across the board: whether it be a hypothetical accident involving gasoline, propane or sulfuric acid, fire chiefs who see a high risk to nearby residents are more likely to favor escorts for shipments of hazardous materials routed in or through their towns than are those who do not.

TABLE 1

Support for Towns Providing Escorts
for Hazmat Shipments
By Perceived Risk of Selected Materials*

(% Supporting Escorts)

	Perceived Risk				
MATERIAL					
Gasoline	High Risk To...		Not High Risk To..		Weighted N	X^2
First Responders	39%	(119)	30%	(50)	169	N.S.
Nearby Residents	44%	(108)	23%	(61)	169	.007
Natural Environ.	37%	(110)	34%	(59)	169	N.S.
Sulfuric Acid						
First Responders	36%	(162)	29%	(7)	169	Inappl.
Nearby Residents	38%	(154)	13%	(15)	169	.055
Natural Environ.	36%	(151)	33%	(18)	169	N.S.
Propane						
First Responders	40%	(148)	10%	(21)	169	.007
Nearby Residents	34%	(145)	17%	(24)	169	.032
Natural Environ.	47%	(90)	24%	(79)	169	.002

* Numbers in parentheses are the base for the computation of percentages. For example, among 108 fire chiefs who perceive spilled gasoline to constitute a high risk to nearby residents, 44% would enable the town to provide escorts for hazmat shipments.

For example, among those who see a high risk to nearby residents from a gasoline spill, 44% support the idea of escorts; among those who do not see a high risk to nearby residents, 23% favor a ban. For a spill involving sulfuric acid, the corresponding percentages are 38% versus 13% respectively, while the comparisons for the scenario entailing a propane spill are 34% versus 17%. At the same time, fire chiefs are especially sensitive to the risks associated with a propane spill: they endorse the idea of escorts if they perceive a high risk to first responders and to the natural environment as well as to nearby residents if this fuel is specified in the scenario.

Rated Effectiveness of the Hazmat Transport System

Several questions were asked concerning the use and effectiveness of various approaches to risk management. [3]

As you may know, the State of Connecticut has adopted the Federal Code regulating shippers and carriers of hazardous materials. Based on your experience, would you say these regulations are effective, somewhat effective, or not effective?

What about state and local inspections of vehicles carrying hazardous materials? Would you say this program is effective, somewhat effective, or not effective?

The regulatory approach has some albeit limited impact in managing this system according to our informants. Only a minority (one in five) consider each of the above features to be "effective". As we have noted elsewhere (Zografos and Warkov, 1990) evaluations of effectiveness may be a function of political orientation and organizational position as well as hazmat transport experience. How fire chiefs perceive the regulatory system may influence their views on other issues such as the role of restricted routing. An index based on these two items was employed to analyze the correlates of perceived effectiveness of the regulatory system. Level of perceived risk to first responders, nearby residents, and the natural environment due to accidents involving gasoline, propane or sulfuric acid is unrelated to how fire chiefs rate the effectiveness of the regulatory system. [4] But, rated effectiveness appears to explain support for restricted

---

[3] The section describing the variables is excerpted from Warkov, 1989:19.

[4] This holds true for both the 2 point and 3 point perceived effectiveness scales.

routing in five of seven instances. These results are shown in Table 2. For example, among those who consider either or both feature of the regulatory system effective (and accordingly are scored 'high') 61% would empower towns to ban the use of roads for specified hazardous materials; while among those rating neither effective, the corresponding percentage is 85. The lower the rated effectiveness score, the more likely the support for proposals to have the town restrict hazmat vehicles to specific routes, prohibit hazmat transport during heavy traffic periods, require permits for shipper and carriers of hazmat, and ban the use of roads and bridges for specific materials.

TABLE 2

Support for Seven Restricted Outing Proposals, by Perceived Effectiveness of the Hazmat Transport System *

Restricted Routing	Perceived Effectiveness**				Weigh-Ted N	X^2
	Not Effective		Effective			
1) Restricted Routing	85%	(115)	61%	(54)	169	.000
2) Peak Period Prohibition	92%	(115)	80%	(54)	169	.019
3) Permits for Shippers	83%	(115)	57%	(54)	169	.000
4) Permits for Carriers	81%	(115)	65%	(54)	169	.023
5) Provide Escorts	36%	(115)	35%	(54)	169	N.S.
6) Lower Speed Limits	70%	(115)	69%	(54)	169	N.S.
7) Ban Use of Roads	72%	(115)	59%	(54)	169	.093

* Numbers in parentheses are the base for the computation of percentages.

** Not effective = negative ratings on both items. Effective = effective or "somewhat effective" on one or both items.

Population Size and Highway Systems

In the section that follows, we consider these results in light of community variation in population size and highway system complexity (i.e., is the town intersected by an expressway and/or other principal arterials?). All things considered, large cities and towns should have complex highway systems and maintain fire departments controlling organizational resources not to be found in small towns. The data (not shown) confirm this pattern: both highway system complexity and town risk assessment activity (specified earlier) correlate with population size. In other words, towns that score high on level of town risk assessment activity are also likely to be (1) big cities; and (2) towns and communities with complex highway systems according to our measures. On the other hand, level of risk assessment activity has nothing to do with fire chief risk perception of the gasoline, propane or sulfuric acid scenarios nor are risk assessment activity scores related to a summary measure of support for empowering towns to impose restricted routing to manage hazmat transport. However, risk perception is related to support for towns to undertake restricted routing.

These results are elaborated employing a General Linear Model (1982) which shows (see Table 3) that both risk perception (beta = .289), perceived effectiveness (beta = .251) and population size (beta = .1871) predict support for restricted routing. The inclusion of our measure of risk assessment activity and highway system complexity add nothing of statistical significance to this model (table not displayed).

TABLE 3

Regression Estimates for Restricted Routing

Predictor Variables	Dependent Variable (RR7)* Beta (Standardized)	Adjusted R^2=.1456 p
Risk 9[1]	.2892	.002
EFFINDEX[2]	-.2511	.010
LPOP[3]	.1871	.054

* RR7 = Summary score on 7 restricted routing proposals.
1. Risk 9 = Summary score on 9 items. See text for risk scenarios.
2. EFFINDEX = Effectiveness index. See Text
3. LPOP = Log, population size, 1980.

Concluding Discussion

Our presentation suggests a few factors that contribute to the views of fire chiefs about hazmat transport issues. Concern about the safety of the public, especially nearby residents who may be exposed to highway spills involving gasoline, propane or sulfuric acid encourages fire chiefs to support more aggressive intervention on the part of local governments. Fire chiefs are also sensitive to the risks associated with all aspects of a propane spill and favor local action if they see a high risk to first responders and to the natural environment in the event of a propane spill. They also favor restricted routing if they deem the current regulatory system to be relatively ineffective and if population size is taken into account. Probably other features of community responsiveness to hazmat emergencies are involved as well (see Warkov, 1990) including local awareness of hazmat storage of facilities, resources for staffing and equipping local emergency response teams, and training programs.

References

SAS/Statistics: Users Guide. 1982. SAS Institute, Carry, N.C.

Warkov, S. 1989. Hazardous Materials Transport Issues in Connecticut: A State-Wide Survey of Fire Chiefs. a Report to Region One, University Transportation Center, U.S. DOT, Cambridge, MA.

Zografos, K. and S. Warkov, "Hazardous materials Transport: Fire Chiefs as Decision-Makers," Transportation Research Record, Transportation Research Board, National Research Council, Washington, D.C., accepted for publication.

Acknowledgement

S. Warkov acknowledges the support of the Federal D.O.T., Region 1, Transportation Centers Programs, Massachusetts Institute of Technology, Cambridge, MA in the conduct of research reported in this paper.

DATABASE DEVELOPMENT

HAZARDOUS MATERIALS DATA: A FEDERAL PERSPECTIVE
Richard C. Hannon[1]

Abstract

This paper discusses the evolution and current status of the Hazardous Materials Incident Report System (HMIRS) maintained by the Research and Special Programs Administration (RSPA) of the U.S. Department of Transportation (DOT). The basis for the Federal collection of information on hazardous materials incidents during transportation is reviewed, and major efforts to improve the collection and use of that data are summarized.

The recent revisions to the Hazard Materials Incident Report (DOT Form 5800.1), used to collect hazardous materials incident data, are reviewed for enhancements to the collected data and the potential application of the expanded data in comparative analyses. Further data enhancements currently being legislatively proposed are briefly discussed, and a set of conclusions are presented as opportunities for future research and analysis.

Introduction

Data Collection and Information Management is only one of five themes this conference, "State and Local Issues In Transportation of Hazardous Materials: Towards A National Strategy," is organized around.

---

[1]Chief, Policy Development and Information Systems Division, Office of Hazardous Materials Transportation, Research and Special Programs Administration, U.S. Department of Transportation, 400 7th Street, S.W., Washington, DC 20590

However, it may be the most important. Data provide information and are the basis upon which the other four themes comprising this conference may be evaluated.

Whether the issue is community preparedness and emergency response, risk evaluation and communication, the consideration of routes or sites, or inspection and enforcement, data must be collected and managed in order that information may be provided to interested parties and decision makers. Without the collection of data and the information it provides, public policies implemented to ensure that hazardous materials are transported safely would, by definition, be made in ignorance.

The Research and Special Programs Administration's (RSPA) Hazardous Materials Information System (HMIS) is the principal source of information on the safety of hazardous materials transportation for the Department of Transportation (DOT). The HMIS actually consists of six subsystems: incidents, exemptions to regulations, interpretations of regulations, container approvals and registrations, inspection and enforcement proceedings, and highway route controlled radioactive material shipments. The incident subsystem, officially known as the Hazardous Materials Incident Reporting System (HMIRS), is the focus of this paper.

Background

Consisting of over 170,000 records, the HMIRS stretches back almost 20 years and represents the best available source of nationwide data involving incidents reported during the transportation of hazardous materials. The varied applications of the HMIRS include research and analysis on the safety performance of the containers and vehicles used to transport hazardous materials, along with assessing the effectiveness of the governing regulations.

RSPA's regulations and definitions are contained in the Code of Federal Regulations, Title 49, Parts 171-179. An incident is reportable to RSPA whenever an unintentional release of a hazardous material takes place during the course of transportation, including loading, unloading, and temporary storage, or whenever a telephonic report is required under RSPA's regulations.

All modes of commercial transportation are covered under RSPA's regulations. The requirement to report incidents involving a hazardous material is not, however, comprehensive within each mode. During transportation by water, only those incidents involving non-bulk shipments of hazardous materials are reportable to RSPA. Releases of hazardous materials involving bulk shipments by water are reportable to the Coast Guard. Highway carriers with interstate operations fall under RSPA's reporting requirements. However, an incident occurring during an intrastate highway shipment by a non-interstate carrier is not required to be reported to RSPA unless the material is a hazardous substance or hazardous waste. This is not the case in the rail mode. Unintentional hazardous material releases that occur on an intrastate rail shipment such as on a short line railroad are required to be reported to RSPA via Hazardous Materials Incident Report (DOT Form 5800.1.)

The hazardous materials incident reporting requirements were established on October 31, 1970, with the passage of the Hazardous Materials Control Act of 1970. The passage of the Hazardous Materials Transportation Act on January 3, 1975, marked the beginning of an upsurge in reported incidents that culminated in the late 1970's with almost 18,000 Hazardous Material Incident Reports being submitted annually. A change in the reporting requirements was made effective January 1, 1981, that excluded consumer commodities, wet electric storage batteries, or paint, enamel, lacquer, stain, shellac, etc., in packaging of five gallons or less. This modification of the requirements does not apply if an incident results in death, injury or property damage over $50,000; the material is being transported by air; or the material is classified as a hazardous waste. After implementation of the revised reporting requirements, the number of recorded incidents remained constant at approximately 6,000 per year, with an increase to 7,481 in 1989 due to an increase in reports received from small package carriers and railroads. The number of reported incidents is presented graphically in Exhibit 1: HMIRS Incidents.

The Hazardous Materials Incident Report form was recently revised to provide more meaningful and comprehensive incident data, especially in terms of incident causation and consequence. Effective January 1, 1990, the expanded and more detailed report form is required for all incidents reported to RSPA.

Beyond the formal rulemakings and public law affecting the reporting of hazardous materials incidents, a noteworthy aspect of RSPA's incident data is that, beginning in 1982, deaths and injuries were verified as having been caused by the hazardous material involved in the incident. The last eight years (1982-1989) have produced what might be broadly described as a consistent block of data containing almost 50,000 incidents upon which analysis may be conducted. An example of summary statistics from these years is presented in Exhibit 2: Incident Statistics by Mode and Reporting Year.

The Revised Incident Report Form

A final rule notice appeared in the Federal Register [54 FR 25808] on June 19, 1989, requiring a revised version of the Hazardous Materials Incident Report, Form 5800.1, be used for all incidents occurring on or after January 1, 1990. RSPA initially proposed changing DOT Form 5800.1 with an advance notice of proposed rulemaking (ANPRM) in the Federal Register [49 FR 10048] on March 16, 1984. A notice of proposed rulemaking (NPRM) appeared in the Federal Register [52 FR 9996] inviting comment on several specific proposed changes to its system of collecting information on incidents involving the transportation of hazardous materials. Extensive improvements and enhancements to the Hazardous Material Incident Report database are expected to result from usage of the revised DOT Form 5800.1 and additional telephonic reporting criteria. Incidents reported telephonically under 49 CFR §171.15 to the National Response Center must also be reported to RSPA on DOT Form 5800.1. The provision for making a telephonic report has been expanded to include circumstances that involve an evacuation of the general public lasting one or more hours, closure of one or more major transportation arteries or facilities or shutdown for one hour or more, or alteration of the operational flight pattern or routine of an aircraft.

Additional changes implemented by the adoption of the new form include an extension of the reporting deadline from 15 days to 30 days after the date of the incident, and an expansion of the package failure designations from 17 to 46 specific codes in five general areas of failure. The new form has also added more data fields and greatly expanded original data fields existing on the earlier form. Examples of improvements in the data fields contained in the new version of 5800.1 that will contribute to improved analysis are: time of incident (although contained on the old form, this field will now be entered electronically into the database); county location (a new field); and a detailed damage breakdown (four specific and one other category of damage). In all, there are 24 data fields that are new or have been enhanced either through electronic retrieval capability or a more detailed description of the information. A summary of the data elements collected in the HMIRS, indicating those elements that have been added or enhanced on the new form, is provided in Exhibit 3: Information Contained in DOT Form 5800.1.

Significance of the New Form

The effect of the changes and enhancements to the new form extends beyond improvement to the database alone. Data captured by the new form will create a much greater level of compatibility with other databases and the improved detail will aid in conducting analysis and contribute to more informed policy development. Common data fields among OHMT's database and the databases of the Federal Railroad Administration (FRA) and the Federal Highway Administration (FHWA) will result in an improved ability to identify occurrences of non- or under-reporting of hazardous material incidents and aid in enforcement actions.

In conjunction with the development of a revised and enhanced form used to gather data, a great deal of effort has been extended to improve the dissemination and accessibility of the incident database. A summary of hazardous materials incident data has been presented in the Annual Report to Congress on Hazardous Materials Transportation and the Transportation Safety Information Report, which is an annual compendium of selected national transportation safety statistics produced by the Transportation Systems Center (TSC).

Government agencies and the private sector also request specific and customized information from the incident database. These requests can result in lengthy printouts containing thousands of records, summary tables, or copies of the actual incident reports. In 1989, RSPA responded to approximately 450 requests for incident data.

A data sharing program has been underway at RSPA since the early 1980's that allows government agencies to directly access the incident database. Currently there are 85 Federal agency accounts and 29 state and local agency accounts. To provide account holders immediate and easy access to the incident database, a menu-driven program was made available in 1989. The menu program allows a user to quickly select parameters, retrieve data, and generate appropriate reports without extensive knowledge of the structure of the incident database.

Incident Reporting: Past Recommendations and Current Effort

The integrity and effectiveness of RSPA's hazardous materials incident database has been reviewed by both DOT and outside agencies. The General Accounting Office in a November 1980 report, Programs for Ensuring the Safe Transportation of Hazardous Materials Need Improvement, cited deficiencies in RSPA's hazardous materials database. The Secretary of Transportation's Safety Review Task Force in their 1985 document titled, Report on the Hazardous Materials Program of the Research and Special Programs Administration, also made a series of recommendations suggesting improvements to RSPA's hazardous materials incident database.

The Office of Technology Assessment (OTA) in July 1986 published a comprehensive study, Transportation of Hazardous Materials, in which a section titled "Data and Information Systems for Hazardous Materials" discussed in detail RSPA's HMIRS database. OTA found that improvements were needed regarding the accuracy and application of the HMIRS. Recently, in November 1989, the GAO reevaluated RSPA's HMIRS database in their report, Railroad Safety: DOT Should Better Manage Its Hazardous Materials Inspection Program, and recommended that RSPA require updated information on incidents, and share accident and enforcement data with other agencies.

RSPA has undertaken a number of initiatives aimed at addressing criticisms expressed in the studies listed above. Staff has been increased to process incident reports, greater analysis of the data is now taking place, and dissemination of the data is improving. Comparison of the HMIRS database with other DOT databases is being established and, in the case of the FRA, rail equipment accident/incident report database cross-checking with the HMIRS is now routine. Also, incidents brought to RSPA's attention through the news media or by one of the modal administrations through a request for which a DOT Form 5800.1 report can not be located are referred for appropriate enforcement action.

Outreach and education campaigns are also being conducted by RSPA to encourage and inform the reporting community about the requirements and procedures associated with submitting hazardous materials incident reports. Approximately 30,000 copies of the new DOT 5800.1 Form and a guide for completing the form were distributed to the reporting community last fall.

In addition, RSPA has joined with the Federal Emergency Management Agency to sponsor and maintain a Hazardous Materials Information Exchange (HMIX), an electronic bulletin board to provide information regarding prevention, preparation and mitigation of hazardous materials emergencies. The HMIX system also provides a bulletin service and nine information conferences that list training events, literature, online databases, regulations, news events, and organizational resources.

Proposed Data Enhancements

The Hazardous Materials Transportation Act (HMTA) is now being discussed for reauthorization by Congress. Currently there are over 20 bills under consideration that address hazardous materials transportation in some manner. Many of these bills contain proposals requiring the collection of additional data or studies necessitating the use of large amounts of sophisticated data.

Some proposals contained in bills that include data provisions involve the designation of routes for transporting hazardous materials and flow studies of certain hazardous materials nationwide or through particular regions. The registration of shippers, the permitting of motor carriers, the identification by shippers of carriers used, along with extending the 49 CFR regulations to intrastate highway shipments of hazardous materials, are also contained in proposed legislation. Highly technical and complex data proposals include real-time satellite tracking, transmission of a manifest to a central reporting system, and the creation of a data center to provide information to emergency responders.

The above listed data proposals are only proposals, contained in bills pending before Congress, of which only a minute portion, even in a modified version, can be expected to pass. It is, however, worthwhile to note the extent and depth of the data provisions contained in the proposed legislation.

RSPA is currently sponsoring a number of studies and research efforts pertaining to hazardous materials transportation data. Recently, RSPA has asked its Transportation Systems Center (TSC) to examine hazardous material transportation incident data collected by representative states and to compare this data with that contained in the HMIRS. The completeness of the states' data will be evaluated along with its integrity and the compatibility with other databases.

Another hazardous materials research effort being supported by RSPA involves identifying those chemicals (estimated at 100 to 500) that constitute 80 percent of the volume of hazardous materials shipped by truck. This project is being undertaken by SRI International of Menlo Park, California. Sandia National Laboratory is also involved in an effort sponsored by RSPA that consists of taking an indepth look at highway routing guidelines and the role of data in making routing decisions.

Summary and Conclusions

Over the past two decades the Hazardous Materials Incident Report database has proven to be a valuable tool in studying and assessing the safety of hazardous materials transportation in the United States. During this period of time, the HMIRS database has been constantly evolving and improving. From the beginning, when only a limited number of HMIRS reports were received each year to the 1980's, when the data became consistent and predictable, the effectiveness of the reporting requirements has been evaluated. Although the HMIRS data has been consistent since 1982 in terms of the number of incidents received, the hazard classes of the materials involved, and the cause and consequences associated with the incidents, an expanded and more detailed report form was needed.

The new form, with its additional data fields represents the potential for much improved research and analysis. The ability to conduct comparative studies with other hazardous materials and non-hazardous materials databases has been greatly increased. State hazardous materials incident databases will now have many more common fields on which to compare data.

Improved analytical capability is not by any means limited to other hazardous materials databases. Accident and incident data collected by other Federal agencies and by state and local agencies represent an extensive set of opportunities.

The revised HMIRS form will provide a much more extensive level of detail that will allow additional analysis to be conducted. The data fields that have been added to DOT Form 5800.1 will contribute a much clearer picture of the factors and resultant consequences associated with hazardous materials accidents or incidents that occur during the course of transportation.

In conjunction with the improvements in the HMIRS form and the advances that have been made in processing, disseminating, and analyzing the data, an effort must be made to obtain more complete and comprehensive submittal of incident reports. The majority of reports received during the first months of 1990 were to some extent incomplete. It is anticipated that with further information and experience the submittal of incomplete forms may be eliminated or at least sharply curtailed and that through greater education and improved enforcement non-reporting by carriers can be reduced. With revisions in the incident report form and greater compatibility with other databases and the potential application of newly developed computerized research tools such as the Census Bureau's TIGER file, the future for research in what has been a commendably safe system of hazardous materials transportation holds great promise.

References

U.S. Department of Transportation, U.S. Government Printing Office, Code of Federal Regulations Title 49: Parts 100 to 177, Washington, D.C., October, 1989.

Annual Report on Hazardous Materials Transportation - Calendar Year 1988, U.S. Department of Transportation, Research and Special Programs Administration, Office of Hazardous Materials Transportation, Washington, D.C., December 1989.

Transportation Safety Information Report - 1988 Annual Summary, U.S. Department of Transportation, Research and Special Programs Administration, Transportation Systems Center, Cambridge, MA., December 1989.

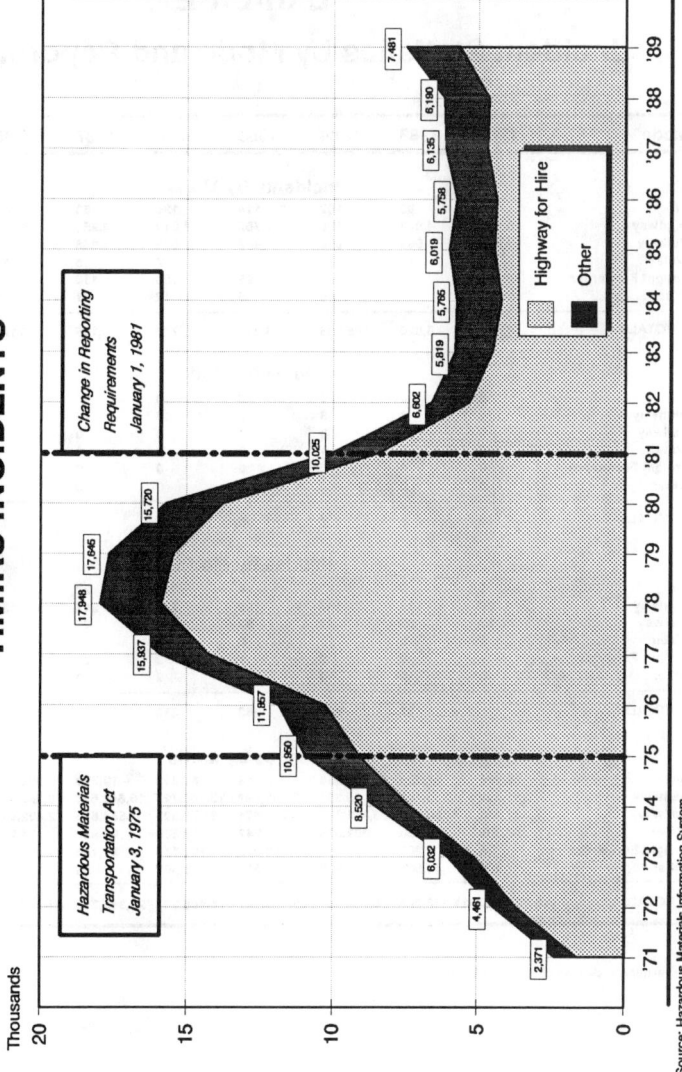

Exhibit 2
Incident Statistics by Mode and Reporting Year

Mode	1982	1983	1984	1985	1986	1987	1988	1989*	Total
Incidents by Mode									
Air	95	66	102	114	120	163	172	187	1,019
Highway	5,662	4,872	4,508	4,752	4,614	4,952	4,904	5,977	40,241
Railway	830	868	996	842	855	886	1,019	1,178	7,474
Water	8	12	8	7	7	15	16	10	83
Freight Forwarder	6	1	145	298	150	118	78	127	923
Other	1	1	6	6	12	1	1	2	30
TOTALS	6,602	5,820	5,765	6,019	5,758	6,135	6,190	7,481	49,770
Deaths by Mode									
Air	0	0	0	0	0	0	0	0	0
Highway	13	8	6	8	16	10	19	8	88
Railway	0	0	0	0	0	0	0	0	0
Water	0	0	1	0	0	0	0	0	1
Freight Forwarder	0	0	0	0	0	0	0	0	0
Other	0	0	0	0	0	0	0	0	0
TOTALS	13	8	7	8	16	10	19	8	89
Injuries by Mode									
Air	0	3	15	4	12	26	6	54	120
Highway	88	118	147	195	229	247	127	205	1,356
Railway	36	68	76	53	59	25	36	36	389
Water	1	0	18	0	2	8	0	7	36
Freight Forwarder	0	0	3	1	12	25	0	15	56
Other	0	0	0	0	2	0	0	0	2
TOTALS	125	189	259	253	316	331	169	317	1,959
Damages by Mode									
Air	26,826	52,525	770,956	12,299	62,813	13,779	562,176	105,011	1,606,385
Highway	11,381,564	9,253,755	11,118,351	12,689,492	13,106,727	15,648,693	18,551,864	15,320,205	107,070,651
Railway	4,331,465	2,559,130	3,353,339	10,273,671	3,077,825	7,554,815	2,432,476	10,265,206	43,847,927
Water	30,000	76,088	509,029	3,242	53,500	99,930	74,262	39,900	885,951
Freight Forwarder	35	300	14,011	13,918	102,117	51,128	15,009	37,655	234,171
Other	200	16,500	975	515	3,385	50	2,700	2,600	26,925
TOTALS	15,770,090	11,958,298	15,766,661	22,993,137	16,406,367	23,368,393	21,638,487	25,770,577	153,672,010

* Preliminary data as of February 27, 1990

Exhibit 3

Information Contained In DOT FORM 5800.1
(Compared to previous edition)

CATEGORY	NEW	ENHANCED	NO CHANGE
▸ Mode of Transportation			★
▸ Date and Time of Incident		★	
▸ Incident Location		★	
▸ Carrier's Name and Address		★	
▸ Carrier's Identification Number	★		
▸ Shipper's Name and Address		★	
▸ Consignee's Name and Address		★	
▸ Shipping Paper or Waybill Identification Number		★	
▸ Material Name and Identification		★	
▸ Hazardous Substance and RQ	★		
▸ Fatalities			★
▸ Injuries		★	
▸ Number of People Evacuated	★		
▸ Dollar Amount of Damage		★	
▸ Consequences of Incident	★		
▸ Type of Vehicle Involved (tank trailer, tank car, etc.)		★	
▸ Transportation Phase (en route, loading, etc.)	★		
▸ Land Use at Incident Site (industrial, commercial, etc.)	★		
▸ Community Type at Site (urban, suburban, rural)	★		
▸ Vehicle Accident or Derailment (estimated speed, highway type, number of lanes)	★		
▸ Packaging Information (type, capacity, etc.)			★
▸ Action Contributing to Packaging Failure	★		
▸ Object Causing Failure	★		
▸ How Package Failed	★		
▸ Package Area that Failed	★		
▸ What Failed on Package	★		
▸ Description of Events		★	

Flows of Hazardous Materials Through States by Rail

Richard Hannon[1] and Paul Zebe[2]

Abstract

This paper presents information on the tonnages of hazardous materials passing through each of the contiguous 48 states (and the District of Columbia) by rail. The quantities are estimated using data from the 1986 Rail Waybill Sample and a BEA-to-BEA rail flows model. Based on the quantity estimates, it appears that most of the hazardous materials traffic moving by rail in a majority of states is passing through, rather than originating or terminating. The states with the most through traffic are (in order) Indiana, Mississippi, Missouri, Illinois, Ohio, Arkansas, Alabama, Georgia, Pennsylvania, and Tennessee. In 27 states and the District of Columbia, 50 percent or more of the traffic, measured in tons, is passing through. In one of these, the District of Columbia, 100 percent of the rail traffic is passing through. In 21 states, less than 50 percent of the hazardous materials traffic moving by rail is passing through. In three of these, Connecticut, Maine, and Rhode Island, there is no hazardous materials through traffic moving by rail.

Introduction

The quantities of hazardous materials being transported are important inputs for emergency planning and response by states and localities. Information on quantities that are originating and terminating within a jurisdiction can generally

---

[1] Chief, Policy Development and Information Systems Division, Office of Hazardous Materials Transportation, Research and Special Programs Administration, U.S. Department of Transportation, 400 7th Street, S.W., Washington, DC 20590.

[2] Economist, Hazardous Materials Special Projects Office, Transportation Systems Center, Research and Special Programs Administration, U.S. Department of Transportation, Kendall Square, Cambridge, MA 02142.

be obtained from those originating or receiving the shipments[3] or from commodity flow databases, such as the ICC waybill database. To get the total picture of hazardous materials movements, however, government agencies concerned with emergency planning and response also need information on the quantities of hazardous materials that are passing through a state or locality. This can be difficult to obtain, since those originating and receiving the shipments can be almost anywhere in the country, and commodity flow databases, such as the ICC waybill database, which generally contain information on originations and terminations, usually have only sketchy and incomplete information, at best, on the routing of shipments.

This paper examines the quantities of hazardous materials passing by rail through each of the 48 contiguous states, along with the District of Columbia. The paper uses data developed for a U.S. Department of Transportation study characterizing the flows of selected hazardous materials by rail. A previous U.S. Department of Transportation study [Maio and Liu, 1987] has addressed hazardous materials through traffic moving by highway. That study, which looked at the situation in 22 regions, each consisting of one or more states, found that, on average, eight percent of the total tonnage of hazardous materials moving by highway was passing through (the percentages for the regions ranged from 0 to nearly 32 percent) [Maio and Liu, 1987, pp. 5-10 to 5-17]. Given the generally short-haul nature of highway movements of freight, it is not surprising to find that only a small proportion of the hazardous materials traffic moving by highway is passing through. Given the more long-haul nature of rail movements of freight, it is not expected that a similar conclusion will be arrived at for rail movements of hazardous materials. Rather, it is expected that a large proportion of the hazardous materials traffic moving by rail will found to be passing through.

Methodology

The quantities of hazardous materials passing by rail through each of the 48

---

[3]Highway data has been obtained by states and others through surveys of vehicles hauling hazardous materials on the highways and questionnaires sent to shippers, carriers, and manufacturers [U.S. Congress, 1986, p. 60]. One state that has employed surveys to obtain highway flow data is Virginia [see Schmidt and Price, 1979]. Other states that have employed surveys include Washington and South Dakota [U.S. Congress, 1986, p. 61]. Questionnaires have been used to gather information on the truck transport of hazardous materials by city of Memphis, among others [U.S. Congress, 1986, p. 61]. The Office of Technology Assessment (OTA) reports that rail data is "generally available on request from the major railroads" [U.S. Congress, 1986, p. 61]. Oregon is one state reported by the OTA to collect information from the railroads on shipments of hazardous materials moving within its borders [U.S. Congress, 1986, p. 62].

states and the District of Columbia were unavailable from any commodity flow database, and, as a consequence, were estimated. This estimation was accomplished using ALK Associates' BEA-to-BEA rail flows model.[4] This model combines a simplified rail network, ALK Associates' "BEA Rail Network", in which the nodes are BEA's and the links are representations of the major rail corridors between the BEA's, with a minimum path assignment algorithm that simulates the most likely route used by the railroads. The model is based on a set of procedures for estimating the flows of commodities by rail developed at Princeton University. The data used as inputs for the model were the tonnages of hazardous materials (i.e., STCC 49 commodities) originating and terminating in each of the BEA's in the contiguous 48 states. These tonnage figures were estimates based on sample tonnages and their associated expansion factors taken from the ICC Carload Waybill Sample for 1986. Traffic terminating in Canada and Mexico, it should be noted, is not included in the Waybill Sample. Consequently, tonnages of hazardous materials going from the U.S. to Canada or Mexico by rail are not included in the estimates that are presented in this paper.

Overview of Hazardous Materials Rail Traffic in 1986

In 1986, a total of 1,476,535,807 tons of freight moved by rail in the U.S. Of this, 62,961,696, or four percent, was hazardous materials. The top five origination states for hazardous materials moving by rail in 1986 were, in order, Texas, Louisiana, Illinois, New York, and Florida. The top five destinations states were, in order, Texas, Illinois, Florida, California, and Louisiana.

Exhibit 1 presents a picture of the flows of hazardous materials moving by rail in 1986. As can be seen in the exhibit, the heaviest flows of hazardous materials by rail are generally from the Gulf Coast states of Texas and Louisiana north, toward the Great Lakes states and also up through the southeastern states to the Northeast. There is also a fairly significant movement west from the Texas Gulf coast toward California, as well as a fairly significant movement of hazardous materials between the Midwest to the Mid-Atlantic coast.

The Quantities of Hazardous Materials Passing Through States

Exhibit 2 presents the estimates of the tons of hazardous materials passing through each of the 48 contiguous states and D.C. for 1986, along with estimates of (1) the tons both originating and terminating in each state, (2) the tons originating but not terminating in each state, (3) the tons terminating but

---

[4] A BEA, or more formally a BEA Economic Area, is an "economic node... that serves as a center of economic activity...and the surrounding counties that are economically related to the center" [U.S. Department of Commerce, 1977, p. 1]. BEA's are defined by the U.S. Department of Commerce's Bureau of Economic Analysis.

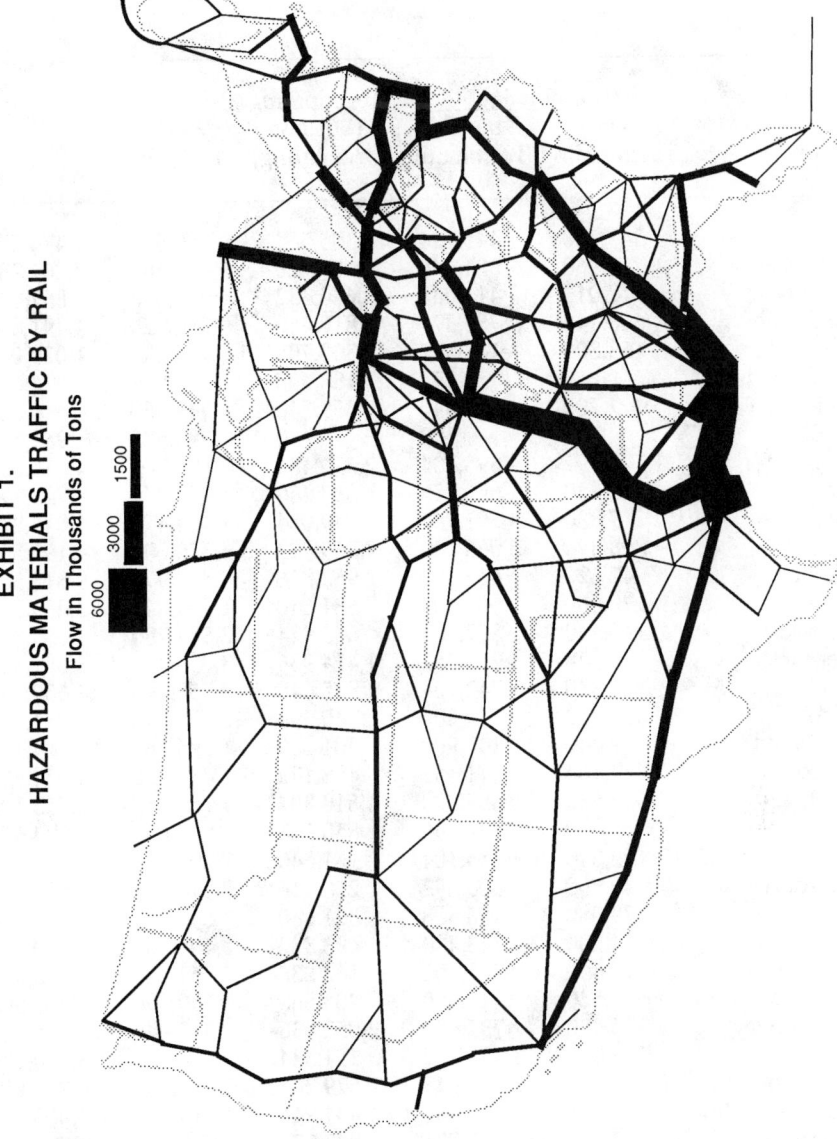

**EXHIBIT 1.
HAZARDOUS MATERIALS TRAFFIC BY RAIL**
Flow in Thousands of Tons

EXHIBIT 2. HAZARDOUS MATERIALS RAIL TRAFFIC BY STATE IN 1986 (IN TONS)

State	Originating and Terminating	Originating but not Terminating	Terminating (but not Originating)	Passing Through	Total
Alabama	343,200	1,445,124	1,305,564	6,380,684	9,474,572
Arkansas	2,960	384,440	464,720	6,415,164	7,267,284
Arizona	17,016	411,060	470,132	1,959,608	2,857,816
California	608,244	786,374	2,810,480	286,720	4,491,818
Colorado	153,800	148,372	300,708	476,204	1,079,084
Connecticut	0	138,272	188,400	0	326,672
Delaware	7,200	598,940	459,560	548,360	1,614,060
D.C.	0	0	0	983,528	983,528
Florida	1,721,397	895,404	2,210,484	87,740	4,915,025
Georgia	321,840	655,716	1,883,460	5,862,757	8,723,773
Idaho	7,880	381,780	249,084	1,403,476	2,042,220
Illinois	308,396	3,065,080	4,269,022	6,550,008	14,192,506
Indiana	42,620	452,112	962,876	8,355,976	9,813,584
Iowa	79,660	503,304	681,640	1,994,642	3,259,246
Kansas	115,440	827,684	603,924	2,887,780	4,434,828
Kentucky	231,560	1,256,352	1,414,936	3,738,472	6,641,320
Louisiana	1,323,320	7,742,104	1,835,648	3,758,264	14,659,336
Maine	631,332	9,600	340,996	0	981,928
Maryland	19,440	147,900	501,600	1,995,180	2,664,120
Massachusetts	7,164	171,152	488,172	485,120	1,151,608
Michigan	26,520	358,800	1,610,304	1,759,572	3,755,196
Minnesota	0	231,896	850,108	892,668	1,974,672
Mississippi	66,840	1,354,104	961,540	7,204,868	9,587,352
Missouri	30,960	606,892	1,228,916	7,013,464	8,880,232
Montana	79,080	155,488	361,680	654,360	1,250,608
Nebraska	34,200	113,480	493,380	1,409,702	2,050,762
Nevada	14,400	241,500	162,288	757,274	1,175,462
New Hampshire	2,520	12,160	205,888	390,296	610,864
New Jersey	326,280	733,680	2,052,436	313,040	3,425,436
New Mexico	119,008	1,474,452	151,344	2,235,316	3,980,120
New York	731,269	2,069,648	679,368	1,356,368	4,836,653
North Carolina	435,604	821,336	1,431,944	1,314,704	4,003,588
North Dakota	43,360	99,736	223,276	723,840	1,090,212
Ohio	245,720	1,722,348	1,845,244	6,478,588	10,291,900
Oklahoma	17,720	658,288	502,912	2,454,492	3,633,412
Oregon	96,560	228,848	540,508	816,460	1,682,376
Pennsylvania	33,700	504,520	1,842,136	5,409,904	7,790,260
Rhode Island	0	0	50,900	0	50,900

not originating in each state, and (4) the total tonnage moving in each state. The total tonnage moving in each state, it should be noted, is the sum of the tonnages in the first four columns of the exhibit. Exhibit 3 presents the states ordered by the tonnages of hazardous materials moving through them by rail.

As can be seen in Exhibit 3, Indiana has the largest amount of hazardous materials passing through it on rail. The other states in the top ten are, in order, Mississippi, Missouri, Illinois, Ohio, Arkansas, Alabama, Georgia, Pennsylvania, and Tennessee. These, and the other states with large amounts of hazardous materials through traffic generally appear to be located on the major routes for the movement of hazardous materials by rail (see Exhibit 1).

Three states, Connecticut, Maine, and Rhode Island, have no hazardous materials passing through them by rail.[5] Other states with relatively small amounts of through tonnage are California, Florida, and New Jersey. The states with relatively small amounts of through tonnage tend to be those at the geographical extremes of the rail distribution system for hazardous materials (see Exhibit 1).

On average, the 48 contiguous states and D.C. have 2,271,374 tons of hazardous materials passing through them by rail. Fourteen of the states have through tonnages that are greater than this, while thirty-four states, plus D.C., have through tonnages that are smaller than the average. The median through tonnage for the states, it might be noted, is 1,356,368 tons.[6]

Exhibit 4 presents the through tonnage by state as a percentage of the total tonnage and Exhibit 5 presents a bar chart with these percentages ordered from the largest to the smallest. The percentage of total tonnage that through traffic represents ranges from 100 percent in the District of Columbia to 0 percent in Connecticut, Maine, and Rhode Island. As can be seen in the exhibits, through traffic makes up 50 percent or more of the total tonnage of hazardous materials moving by rail in 27 states and D.C., and less than 50 percent in the remaining 21 states. On average, through tonnage makes up 51 percent of the total tonnage of hazardous materials moving by rail in a state.

Exhibit 6 presents the tonnage passing through each state per ton originating and/or terminating and Exhibit 7 presents a bar chart with these tonnage ratios

---

[5]Maine, it should be noted, may have hazardous materials traffic passing through it by rail that (1) originates in the U.S. and terminates in Canada or (2) both originates and terminates in Canada. U.S. traffic that terminates in Canada is not included in the Rail Waybill Sample, and neither is traffic that is purely Canadian.

[6]The median is the value above and below which lie an equal number of observations. It, like the average, is a measure of the central tendency of a set of observations.

EXHIBIT 2. HAZARDOUS MATERIALS RAIL TRAFFIC BY STATE IN 1986 (IN TONS) (CONTINUED)

State	Originating and Terminating	Originating but not Terminating	Terminating (but not Originating)	Passing Through	Total
South Carolina	222,812	210,160	1,340,804	1,848,804	3,622,580
South Dakota	0	0	52,400	214,880	267,280
Tennessee	272,264	1,616,000	1,930,992	4,901,384	8,720,640
Texas	4,115,628	8,311,356	4,120,528	2,078,188	18,625,700
Utah	2,604	176,340	534,720	742,742	1,456,406
Vermont	0	0	801,848	706,668	1,508,516
Virginia	51,360	157,040	1,360,628	1,506,228	3,075,256
Washington	469,040	559,080	914,856	596,320	2,539,296
West Virginia	162,020	1,258,948	719,920	1,197,284	3,338,172
Wisconsin	43,400	126,540	470,320	1,134,740	1,775,000
Wyoming	27,200	310,380	310,596	1,015,490	1,663,666

EXHIBIT 3. STATES ORDERED BY THE TONNAGE OF HAZARDOUS MATERIALS PASSING THROUGH THEM BY RAIL

State	Tonnage	State	Tonnage
Indiana	8,355,976	North Carolina	1,314,704
Mississippi	7,204,868	West Virginia	1,197,284
Missouri	7,013,464	Wisconsin	1,134,740
Illinois	6,550,008	Wyoming	1,015,490
Ohio	6,478,588	D.C.	983,528
Arkansas	6,415,164	Minnesota	892,668
Alabama	6,380,684	Oregon	816,460
Georgia	5,862,757	Nevada	757,274
Pennsylvania	5,409,904	Utah	742,742
Tennessee	4,901,384	North Dakota	723,840
Louisiana	3,758,264	Vermont	706,668
Kentucky	3,738,472	Montana	654,360
Kansas	2,887,780	Washington	596,320
Oklahoma	2,454,492	Delaware	548,360
New Mexico	2,235,316	Massachusetts	485,120
Texas	2,078,188	Colorado	476,204
Maryland	1,995,180	New Hampshire	390,296
Iowa	1,994,642	New Jersey	313,040
Arizona	1,959,608	California	286,720
South Carolina	1,848,804	South Dakota	214,880
Michigan	1,759,572	Florida	87,740
Virginia	1,506,228	Connecticut	0
Nebraska	1,409,702	Maine	0
Idaho	1,403,476	Rhode Island	0
New York	1,356,368		

EXHIBIT 4. THROUGH TONNAGE AS A PERCENTAGE OF TOTAL BY STATE

State	Percent	State	Percent
Alabama	67	Nebraska	69
Arkansas	88	Nevada	64
Arizona	69	New Hampshire	64
California	6	New Jersey	9
Colorado	44	New Mexico	56
Connecticut	0	New York	28
Delaware	34	North Carolina	33
D.C.	100	North Dakota	66
Florida	2	Ohio	63
Georgia	67	Oklahoma	68
Idaho	69	Oregon	49
Illinois	46	Pennsylvania	69
Indiana	85	Rhode Island	0
Iowa	61	South Carolina	51
Kansas	65	South Dakota	80
Kentucky	56	Tennessee	56
Louisiana	26	Texas	11
Maine	0	Utah	51
Maryland	75	Vermont	47
Massachusetts	42	Virginia	49
Michigan	47	Washington	23
Minnesota	45	West Virginia	36
Mississippi	75	Wisconsin	64
Missouri	79	Wyoming	61
Montana	52		

EXHIBIT 5.
STATES ORDERED BY PERCENTAGE OF HAZARDOUS MATERIALS TRAFFIC BY RAIL THAT IS PASSING THROUGH STATE

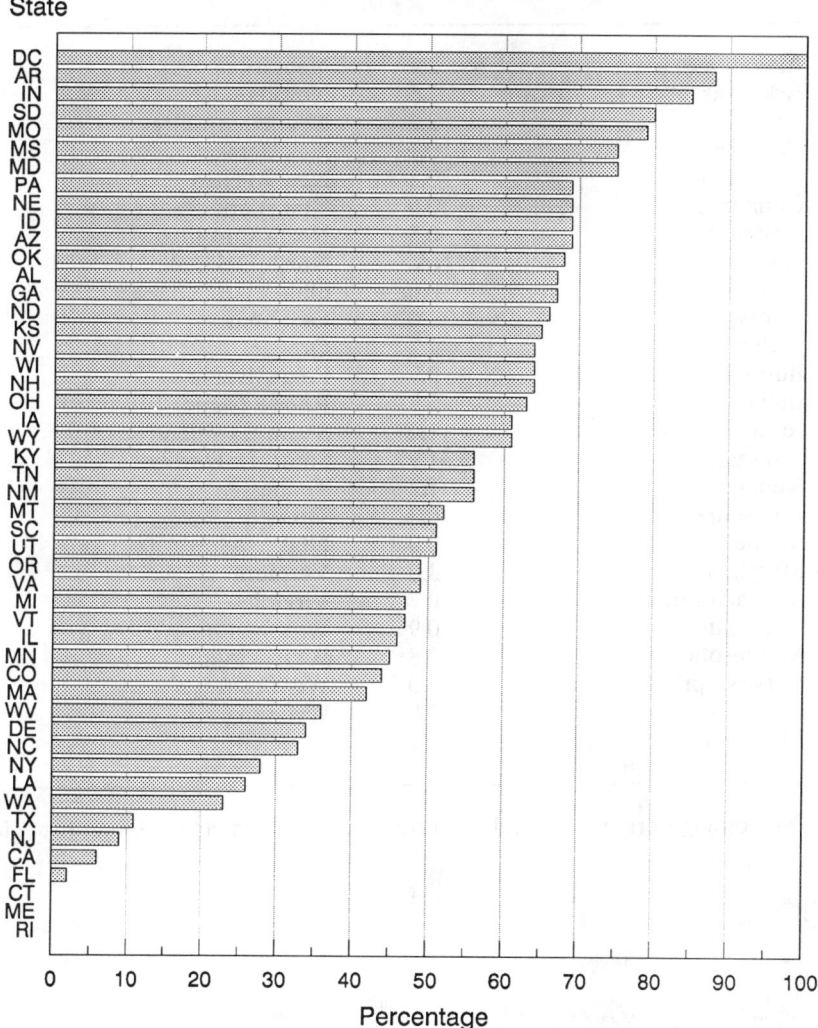

EXHIBIT 6. RATIO OF TONS PASSING THROUGH STATE TO TONS ORIGINATING AND/OR TERMINATING IN STATE

State	Ratio	State	Ratio
Alabama	2.1	Nebraska	2.2
Arkansas	7.5	Nevada	1.8
Arizona	2.2	New Hampshire	1.8
California	0.1	New Jersey	0.1
Colorado	0.8	New Mexico	1.3
Connecticut	0.0	New York	0.4
Delaware	0.5	North Carolina	0.5
D.C.	NA*	North Dakota	2.0
Florida	0.0	Ohio	1.7
Georgia	2.0	Oklahoma	2.1
Idaho	2.2	Oregon	0.9
Illinois	0.9	Pennsylvania	2.3
Indiana	5.7	Rhode Island	0.0
Iowa	1.6	South Carolina	1.0
Kansas	1.9	South Dakota	4.1
Kentucky	1.3	Tennessee	1.3
Louisiana	0.3	Texas	0.1
Maine	0.0	Utah	1.0
Maryland	3.0	Vermont	0.9
Massachusetts	0.7	Virginia	1.0
Michigan	0.9	Washington	0.3
Minnesota	0.8	West Virginia	0.6
Mississippi	3.0	Wisconsin	1.8
Missouri	3.8	Wyoming	1.6
Montana	1.1		

*No tonnage originating and/or terminating; all tonnage passing through.

EXHIBIT 7.
STATES ORDERED BY THROUGH TONNAGE OF HAZARDOUS MATERIALS PER TON ORIGINATING AND/OR TERMINATING

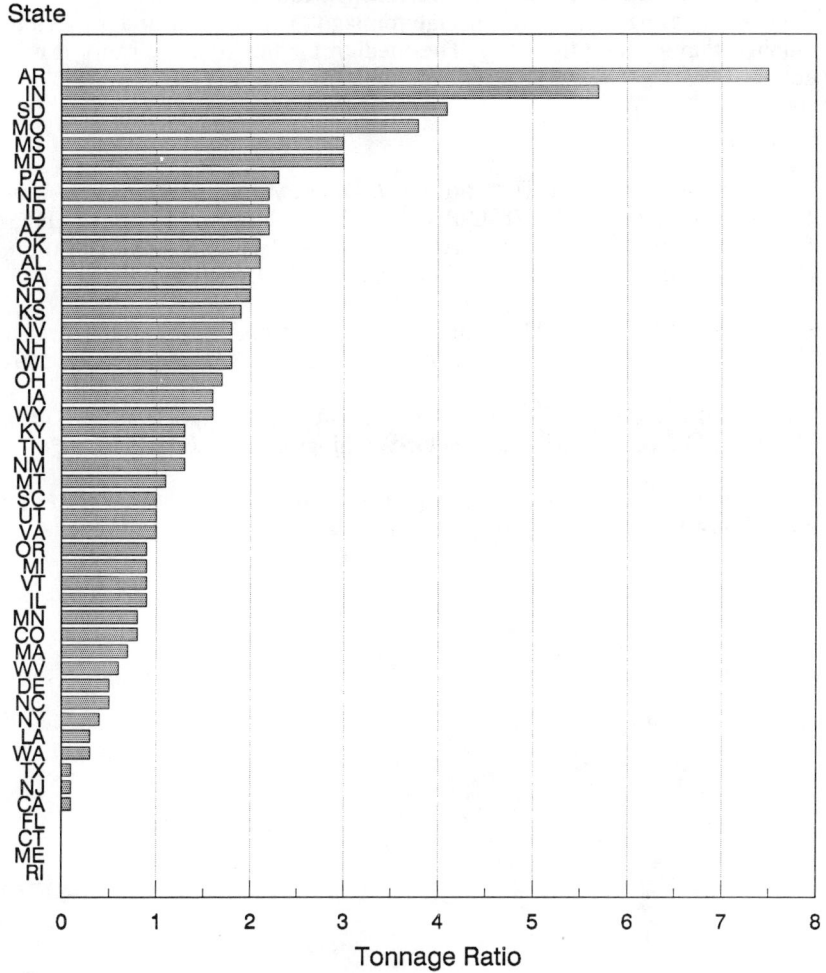

Note: DC not included in bar graph. All tonnage in DC passing through; no tonnage originating and/or terminating

ranked from the largest to the smallest. This tonnage ratio ranges from infinity in the District of Columbia, which has no tonnage originating or terminating, to 0 in Connecticut, Maine, and Rhode Island, which have no through traffic. On average, a state has 1.5 tons passing through for every ton originating and/or terminating (the value for D.C. was not used in the calculation of this average). A total of 21 states have through tonnages greater than this. The other 27 states have ratios of through tonnage to originating and/or terminating tonnage that are less than 1.5. The median for this tonnage ratio, it might be noted, is 1.3.

References

Maio, Domenic, and Liu, Tai-Kuo, *Truck Transportation of Hazardous Materials: A National Overview*, DOT-TSC-RSPA-87-8, Final Report, Office of Hazardous Materials Transportation, Research and Special Programs Administration, U.S. Department of Transportation, Washington, DC, December 1987.

Schmidt, J., and Price, D., "The Flow of Hazardous Materials on Virginia Highways," *Journal of Safety Research*, Vol. 11, No. 3, Fall 1979, pp. 109-114.

U.S. Congress, Office of Technology Assessment, *Transportation of Hazardous Materials*, OTA-SET-304, U.S. GPO, Washington, DC, July 1986.

U.S. Department of Commerce, Bureau of Economic Analysis, *BEA Economic Areas*, U.S. DOC, Washington, D.C., 1977.

ISSUES IN DEVELOPING ARIZONA'S HAZMAT INCIDENT REPORTING PROGRAM
KAREN S. HUBBARD[1]

ABSTRACT

The Arizona Division of Emergency Services (ADES), in cooperation with Arizona State University (ASU) - Center for Advanced Research in Transportation, and the Arizona Department of Transportation (ADOT) are currently developing a statewide post-incident hazardous materials reporting system. This paper describes issues surrounding the development, and adoption of the system, and the methods by which voluntary participation in the system is being solicited.

INTRODUCTION

U.S. Department of Transportation (USDOT) statistics indicate that between 1971-1987 there were 2,305 reported hazardous materials transportation incidents in Arizona. These incidents resulted in 20 deaths, 146 injuries, and $5,727,000 in property damage. This incident information was provided to USDOT by carriers of hazardous materials in accordance with federal reporting requirements. Other federal agencies such as the U.S. Environmental Protection Agency (USEPA) and the Federal Railroad Administration (FRA) require incident reporting based on a standard set of criteria. This information, however; is provided by owners/operators and/or responsible parties and does not usually contain information concerning government response activities.

---

[1]Technical Support Manager, Hazardous Materials, Arizona Division of Emergency Services, 5636 E. McDowell Road, Phoenix, AZ 85008

Currently there is no statewide reporting structure to collect information on hazardous materials incidents in Arizona. Incident information is gathered by several organizations, but a standard format is not used and data is not normally shared with other agencies. Therefore, state and local agencies are unable to analyze the hazardous materials issues, determine risk and appropriate response capabilities, project budgetary requirements to meet identified needs, or provide assistance based upon analysis of past incidents.

Additionally, the statewide Hazard Analysis contained in the January 1989 State Hazardous Material Plan is incomplete in scope because standardized data on incident location, and frequency and probability of occurrence was not available. The Hazard Analysis currently only identifies the locations of fixed-site locations that use or produce hazardous materials, and route locations and modes used for the transportation of hazardous materials.

INCIDENT REPORT SYSTEM DEVELOPMENT HISTORY

In order to provide a means for collecting uniform incident data, ASU, in cooperations with ADOT and ADES, designed and developed a hazardous materials incident report form in 1987. A committee of 12 response agencies reviewed the instrument and organized a pilot study. A four-month pilot test of the form was conducted from December 1, 1987 to April 1, 1988 in Maricopa and Coconino counties.[1] Unfortunately, two major response agencies withdrew their participation midway through the pilot test period. The two agencies concluded that the form added extra paperwork to their workload, and they did not have the time to fill out the forms. Other agencies objected to the form because it did not contain all the data elements needed for their internal use, while other agencies viewed the report form as additional paperwork. The most obvious finding of the pilot test was that the form was considered unsatisfactory to some degree by all of the participating agencies.

INCIDENT REPORT SYSTEM REVISION

A study group was formed in 1989 to review the findings of the pilot test. The following issues are currently being addressed based on the study group recommendations:

- Several incident report forms, including the Arizona Pilot Form and those developed by the States of California and Illinois were reviewed for possible use by the State of Arizona. The California Hazardous Materials Incident Report Form was selected as the most comprehensive and compatible form and modified to reflect the individual and collective data needs of all Arizona agencies. The form also contains all reporting elements of the proposed National Fire Incident Reporting System (NFIRS) HAZMAT Report. See Figure 1.

- A functional definition for a reportable incident was developed, and parallel those contained in the proposed NFIRS HAZMAT reporting system. The definition of a reportable incident varies statewide; i.e., some agencies report cancellations enroute as an incident. Initial estimates conclude that the total number of reportable incidents per year would be approximately 600, with the majority occurring in Maricopa and Pima counties. The incidents could occur at fixed sites or during the transportation or use of hazardous materials.

- A statewide mailing to over 300 fire departments, Sheriff's offices, Local Emergency Planning Committees (LEPCs), emergency management agencies, and state hazardous materials response agencies was made by ADES in March 1990, to solicit comments and commitments of participation in the reporting system. A survey form was developed and included in the mailing to be completed and returned to ADES.

- Responses to the survey have been slow. As of April 27, 1990, a total of 63 agencies have responded to the survey, and 58 have voluntarily agreed to participate in the reporting system. The Arizona State Fire Marshal has agreed to assist ADES in the survey by contacting agencies who do not respond. The regional Fire Marshal representatives will urge all non-responsive agencies to complete and return the survey forms, so that an accurate analysis can be made.

- In order to keep cost to response agencies at a minimum, ADES will supply incident report forms and become the repository for all completed report forms. The reporting will be done on a voluntary basis. ADES will maintain a computer data base for reports using dBase III.

ARIZONA HAZARDOUS MATERIAL INCIDENT REPORT

A	AGENCY NAME	AGENCY ID NO.	AGENCY INCIDENT NO.	AGENCY PHONE NO. ( ) -	ADES CONTROL NO.
B	INCIDENT DATE (MM)__(DD)__(YY)__	TIME NOTIFIED (2400 HR)		TIME COMPLETE (2400 HR)	
C	INCIDENT ADDRESS/LOCATION			CITY	COUNTY

D	WEATHER (Check best descriptor/s) 1__CLEAR 5__HAIL 8__HIGH WIND 3__RAIN 6__ELEC.STORM 9__OTHER 4__SNOW 7__FOG 0__UNKNOWN ESTIMATED TEMP___(Deg F)	PROPERTY USE (USE CODES ON REVERSE) PROPERTY USE ___ ___ SURROUNDING AREA ___ ___ PROPERTY MANAGEMENT __FEDERAL __STATE __COUNTY __CITY __PRIVATE __UNKNOWN

E	RELEASE FACTORS (Check best descriptor/s) 11__INTENTIONAL ACT 21__SUSPICIOUS ACT 30__FAILURE TO CONTROL HAZMAT 31__ABANDONED 40__MISUSE OF HAZMAT 50__MECHANICAL FAILURE 60__DESIGN,CONSTRUCTION, INSTALLATION DEFICIENCY	70__OPERATIONAL DEFICIENCY 71__COLLISION/OVERTURN 80__NATURAL CONDITION 90__UNKNOWN 94__FIRE/EXPLOSION 98__NO RELEASE 99__OTHER________	TYPE OF EQUIP.INVOLVED 10__HEATING SYSTEMS 30__AIR CONDITIONING/REFRIG. 77__CHEM. PROCESSING EQUIP 78__WASTE RECOVERY EQUIP 96__HAZMAT TRANSFER EQUIP 98__NO EQUIP. INVOLVED 99__OTHER________	MOBILE PROPERTY TYPE 10__PASSENGER VEH/ROAD 20__FREIGHT VEH/ROAD 30__RAIL TRANSPORT VEHICLE 40__WATER TRANS VESSEL 50__AIR TRANSPORT VESSEL 60__HEAVY EQUIP-INDUST/AGRI 99__OTHER________
F	ACTIONS TAKEN (Check one or more) 31__RESCUE, REMOVE FROM HARM 32__EXTRICATION, DISENTANGLEMENT 33__EMERGENCY MEDICAL SERVICES 35__SEARCH 36__TRANSPORT	41__REMOVE HAZARD 42__ID/ANALYSIS OF HAZMAT 43__EVACUATION 44__ESTABLISH SAFE AREA 45__MONITOR 46__DECON-PERSON/EQUIP	47__DECON-AREA 61__CROWD CONTROL 62__TRAFFIC CONTROL 63__NOTIFY OTHER AGENCY 64__PROVIDE PUBLIC INFO 71__INVESITGATE	73__SHUT DOWN SYSTEM 82__SECURE PROPERTY 92__REFER TO PROP. AUTHORITY 98__NO ACTION TAKEN 99__OTHER________

G	CHEMICAL OR TRADE NAME (Print or Type)			DOT ID No. / DOT HAZARDOUS CLASS ___ / ___	CAS No. ________
	PHYSICAL STATE STORED 1__SOLID 2__LIQUID 3__GAS		PHYSICAL STATE RELEASED 1__SOLID 2__LIQUID 3__GAS	QUANTITY RELEASED ________ 1) lbs 2) gal. 3) cu ft.	TYPE OF ENVIRONMENTAL CONTAMINATION 1__AIR 3__WATER 7__GROUND 9__OTHER____
	CONTAINER DESCRIPTION 1__FIXED 3__MOBILE 2__PORTABLE	1__ARMORED 1__INSULATED 1__PRESSURIZED	(USE CODES ON REVERSE) CONTAINER MATERIAL__ LEVEL OF CONTAINER___ CONTAINER TYPE___	CONTAINER CAPACITY ________ 1) lbs. 2) gal 3) cu ft	(USE CODES ON REVERSE) EXTENT OF RELEASE ___
	CHEMICAL OR TRADE NAME (Print or Type)			DOT ID No. / DOT HAZARDOUS CLASS ___ / ___	CAS No. ________
	PHYSICAL STATE STORED 1__SOLID 2__LIQUID 3__GAS		PHYSICAL STATE RELEASED 1__SOLID 2__LIQUID 3__GAS	QUANTITY RELEASED ________ 1) lbs. 2) gal. 3) cu ft.	TYPE OF ENVIRONMENTAL CONTAMINATION 1__AIR 3__WATER 7__GROUND 9__OTHER____
	CONTAINER DESCRIPTION 1__FIXED 3__MOBILE 2__PORTABLE	1__ARMORED 1__INSULATED 1__PRESSURIZED	(USE CODES ON REVERSE) CONTAINER MATERIAL__ LEVEL OF CONTAINER___ CONTAINER TYPE___	CONTAINER CAPACITY ________ 1) lbs. 2) gal. 3) cu ft.	(USE CODES ON REVERSE) EXTENT OF RELEASE ___
	CHEMICAL OR TRADE NAME (Print or Type)			DOT ID No. / DOT HAZARDOUS CLASS ___ / ___	CAS No. ________
	PHYSICAL STATE STORED 1__SOLID 2__LIQUID 3__GAS		PHYSICAL STATE RELEASED 1__SOLID 2__LIQUID 3__GAS	QUANTITY RELEASED ________ 1) lbs. 2) gal. 3) cu ft.	TYPE OF ENVIRONMENTAL CONTAMINATION 1__AIR 3__WATER 7__GROUND 9__OTHER____
	CONTAINER DESCRIPTION 1__FIXED 3__MOBILE 2__PORTABLE	1__ARMORED 1__INSULATED 1__PRESSURIZED	(USE CODES ON REVERSE) CONTAINER MATERIAL__ LEVEL OF CONTAINER___ CONTAINER TYPE___	CONTAINER CAPACITY ________ 1) lbs 2) gal 3) cu ft.	(USE CODES ON REVERSE) EXTENT OF RELEASE ___

H	SPECIAL STUDIES	1. a b c d ___ 2. a b c d ___ 3. a b c d ___ 4. a b c d ___
I	HAZMAT IDENTIFICATION SOURCES (Check best descriptor/s) 15__ON-SITE FIRE SERVICES 25__OFF-SITE FIRE SERVICES 60__OFF-SITE NON-FIRE SERVICES 54__CHEMIST 58__TOX CENTER 71__DOT MATERIAL 73__MSDS 75__PLACARDS/SIGNS 78__SHIPPING PAPERS 86__CONTRACT INFO SOURCES 87__COMPUTER SOFTWARE 99__OTHER________	HAZMAT CASUALTIES No. of Contaminated / No. of Injuries / No. of Fatalities EMERGENCY PERSONNEL ___ ___ ___ OTHERS ___ ___ ___

J	VEHICLE MAKE/YEAR	STATE	VEHICLE LICENSE NUMBER	VEHICLE ID NO (VIN)	ICC/DOT/PUC NO / COMPANY NAME
K	REPORTING OFFICER NAME/ID NO (Print or type)			DATE	ADDITIONAL COMMENTS ON BACK YES NO

Figure 1. Arizona Hazardous Materials Incident Report

CODES CODES CODES CODES CODES CODES CODES CODES CODES

PROPERTY USE and SURROUNDING AREA TYPE			CONTAINER MATERIAL	
100 Public assembly	700 Manufacturing	946 Lake/Pond/River	1	Iron and iron alloys
200 Educational	762 Hazmat chem mfg	950 Railroad	2	Aluminum and aluminum alloys
300 Health care	767 Petroleum refinery	961 Freeway	3	Copper and copper alloys
400 Residential	800 Storage	962 County/City road	4	Plastic (includes fiberglass), rigid
500 Mercantile, Business	936 Vacant lot	963 Private road	5	Plastic, flexible
600 Industrial, Utility	941 Open Sea	099 Other - explain in	6	Wood, paper, & cellulose products
650 Agricultural	942 Harbor/Port	comments	7	Glass
			8	NO CONTAINER
			9	Other-explain in comments
			0	Unknown

LEVEL OF CONTAINER	CONTAINER TYPE	EXTENT OF RELEASE
11 Ground Level	01 Tank	1 Confined to vehicle/equipment
10 Above Ground	02 Drum/Barrel	3 Confined to room of origin
40 Below Ground	03 Cylinder	4 Confined to floor of origin
	04 Can/Bottle	5 Confined to structure or origin
	05 Carboy	6 Confined to property use of origin
	06 Boxes/Cartons	7 Release beyond property use of origin
	07 Bags	8 NO RELEASE
	08 Sump/Pit/Pond	9 Other - explain in comments
	09 Well	0 Unknown
	10 Machinery/Processing Equipment	
	11 Pipe	
	18 NO CONTAINER	
	19 Other - explain in comments	

COMMENTS:
_____
_____
_____
_____

IMPORTANT INSTRUCTIONS:

CHANGE: If the information on a previously submitted form needs to be changed mark the CHANGE box and submit form with the correct information.

DELETE: If a certain report needs to be deleted from the database mark the DELETE box, complete sections A,B,C, and L, and submit form.

NOTE: *ALL THE SHADED SECTIONS MUST BE COMPLETED.*
SECTION:
- A Enter all information about the reporting agency. ADES Control No. is assigned when making phone notification to ADES. Phone number 231-6346.
- B Enter the date (month,day,year), notification and completion time of the incident (use 2400 time clock time).
- C Enter the location or address of the incident (include city & county).
- D Check the appropriate weather descriptor(s) at the time of the incident & indicate the approximate temperature in ° F. Enter property use and surrounding area code(s) as appropriate. Indicate the agency responsible for property management.
- E Check the item(s) that describe(s) the cause of the incident, the type of equipment involved in the incident, and the mobile property type if any.
- F Check the item(s) that indicate(s) which action(s) you took as a responder to the incident.
- G List the chemical or the trade name(s) of the hazardous material(s) involved in the incident. Include information required in the boxes. Check the information in the boxes that describes the hazardous material. Use the appropriate codes for Container Type, Container Material, and Extent of Release.
- H This section indicates special studies. Leave blank unless otherwise indicated.
- I Check the item(s) that describe(s) how the material was identified.
- J If vehicle/mobile property was involved in the incident enter information about that vehicle.
- K Print your full name or your ID number and enter the date of report. Indicate (Y/N) whether there are additional comments.

Figure 2. Arizona Hazardous Materials Incident Report

- Incident Commanders will be responsible for completing the Incident Report Form and submitting it to ADES. A control number for each incident will be assigned by ADES to avoid duplication in reporting.

- Telephone reporting through an 800 toll free number, as an option to hardcopy submittal, was explored. It was determined that equipment, maintenance, and associated overhead costs are not currently budgeted and would add an additional financial burden to the ADES.

- A summary analysis of incident information will be shared with all participating agencies by ADES on a quarterly basis. Agencies with modems will be able to access the information by computer during normal business hours.

- Statewide participation will continue to be irregular under a voluntary submittal program. The possibility of creating new legislation mandating reporting will be explored. Mandatory reporting legislation currently exists in the State of California.

SUMMARY

Several political and regulatory issues can be addressed by the adoption of the incident reporting system:

- Newly adopted Occupational Safety and Health Administration (OSHA) Regulation 1919.120, which was effective March 6, 1990, will force emergency response agencies to study and review the appropriate level of hazardous materials response that can reasonably be delivered to their community. A key element in the decisionmaking process will be the availability of information pertaining to the type, magnitude, location, and probability of incident occurrence.

- Section 303(b) of Title III, the Emergency Planning and Community Right-to-Know Act of 1986, requires LEPCs "....to evaluate the need for resources necessary to develop, implement, and exercise the emergency plan, and make recommendations with respect to additional resources that may be required, and the means for providing such additional resources." The availability of structured incident information will provide LEPCs with valuable data that may be used to develop, project, and forecast hazard and risk analyses, as well as recommend appropriate response resources for their jurisdiction.

The absence of accurate hazardous materials incident data will continue to hamper response and recovery program planning projections. The adoption of a statewide reporting system in Arizona will contribute toward the development of an accurate assessment of the State's hazards and needs; thereby benefiting both citizens and responders. ADES, in cooperation with ASU and ADOT will continue efforts toward the establishment of this reporting system.

REFERENCES

1. Radwan, Essam and Field, "Keeping Tabs on Toxic Spills," Civil Engineering, April 1990.

Developing a HazMat Incident Evaluation Program
by
Dan L. Williams[1]
Francis Kaikumba, Ph.D.[2]

Introduction

All communities are at risk of experiencing a hazardous materials incident, whether in transportation or at a fixed facility. However, many communities are unprepared to cope with this emergency. Administrators of state agencies and local response organizations often face fierce competition for tax dollars in support of their hazardous materials emergency response programs. Historically, little statistical information has been available to support their call for additional budget dollars in the areas of personnel, training, and equipment. It is difficult to plan for an emergency when you have little understanding of the risks, the frequency of incidents, the type of chemicals generally involved, and other vital information. The Illinois Hazardous Materials Incident Evaluation Program was designed to address the state and local needs, and position government and industry in a more responsive and proactive mode. The result is a program that works.

Historical Review

In 1985, the Illinois General Assembly passed Public Act 83-1368 which expanded the membership and responsibilities of the state's Hazardous Materials Advisory Board (the Board). It now has as its members, the directors of 12 state agencies, representation from the major statewide response organizations (Fire Chiefs' Association, Chiefs' of Police Association, Sheriffs' Association, and Emergency Services Management Association), and 4 individuals appointed by the Governor (representatives from a volunteer fire department, the insurance industry, and 2 members from the business community). This unique

---

[1] Deputy Director, Office of the State Fire Marshal, 1035 Stevenson Drive, Springfield, Illinois 62703.

[2] Programs Manager, Illinois Department of Transportation, 420 West Washington Street, Springfield, Illinois 62701.

membership allows for high-level discussions of key issues confronting the state. Board members are of sufficient rank that they can commit the resources of their organization and make major policy decisions. The Board has met regularly since 1985, and has established several standing committees. During this period, significant recommendations were offered to the Governor, legislature, and state and local governmental bodies. These included, among others: developing countywide hazardous materials plans (more than a year before enactment of SARA Title III), creating a statewide hazardous materials plan, implementing a statewide incident evaluation program, pre-positioning hazardous materials emergency response supplies and equipment, and other innovative programs.

During the mid 1980s, it became clear that state and local governments had little information about the frequency and consequences of hazardous materials incidents. Consequently, state and local planning needs were not thoroughly addressed in some cases. More detailed information was needed.

Developmental Strategies

The last thing government agencies or private companies wanted to hear was "we have another form for you to submit." The Board realized that there had to be strong incentives for participation in a voluntary program. It felt that there was a pressing need for more information, and that an incident reporting system would meet this need. One incentive was sharing information obtained to aid cities in justifying their request for sustained or increased dollars for hazardous materials planning, training, and response. An annual report is now published and widely distributed. The report reviews recent developments in the hazardous materials arena, overviews data received, and offers analyses. Another strategy was to make the form simple, keeping it to one page, and minimizing the opportunities for reporting errors. The Board decided on a series of questions that generally featured limited options to each question versus "a fill in the blank" type of response. This assisted in data input and data query.

To develop the form, a special committee was formed. Its challenge was to develop a standardized incident report and evaluation form for use throughout Illinois. The committee had representation from many diverse groups, including the emergency response communities. The committee met for many months and worked through several drafts before giving final approval to the forms. The committee then presented its recommended forms and an instruction manual to the full Board. Both were approved and are now in effect.

A key goal of the program is to identify the problem areas and target resources accordingly.

Implementation

Once the mechanics of the process had been resolved, an implemental plan was needed. It is important to note that there was no specific appropriation for this program. State agencies and statewide organizations agreed to absorb the costs of postage to send forms and manuals to nearly 3,000 city and county organizations. Some agencies combined this special mailing with routine mail, thus maximizing expenditures of funds. It was determined that local fire departments would more readily accept the program if forms and manuals were sent with a cover letter by the State Fire Marshal's Office. Similarly, the police departments were more willing to participate if materials were sent from their parent organization. As lead agency to the Board, IDOT absorbed staff costs and the printing expenses.

In addition to direct mailings, the incident evaluation program was the topic of discussion at numerous conferences. The Board sponsors an annual hazardous materials conference; the weekend conference generally attracts 350 to 400 planners and responders. The Board also conducted a series of regional conferences in 1987/1988 to discuss the entire gamut of new programs (state hazardous materials plan, countywide plans, regional response teams, hazardous materials training program, incident evaluation program, and others). Similarly, the matter was also presented at various conferences conducted by other organizations, i.e., Fire Chiefs' Association, Association of Fire Protection Districts, Local Law Enforcement Police Officers Training Board, and other groups. Further, the incident evaluation program was made a part of the course curriculum for the eight-hour hazardous materials basic awareness course (taken by over 10,000 state and local responders since 1987).

The Board recommended that the Illinois Department of Transportation's Hazardous Materials Section be given the responsibility of managing the data collection process, including data input and analyses of the data. Results are published in the Board's Annual Reports. Local governments and state agencies can also obtain specific data on an as-needed basis from IDOT.

IDOT manages the system using a mainframe computer. The program was written in NOMAD II language which allows the user to interact with the data base and update records immediately.

Combined Forms

A two-part form was developed to standardize reporting from police departments, the fire service, emergency services offices, industry, and other reporting organizations. The key challenge was to obtain information from diverse organizations that would serve all interests as a whole. Numerous organizations have major hazardous materials responsibilities. For example, the state transportation department and local highway departments erect barricades, transport lime and sand to hazardous materials sites, and assist in other ways during an emergency. The state Commerce Commission has responsibility for enforcing the rail transportation regulations and responding to derailments. The IEPA, Public Health Department, Department of Nuclear Safety, Agriculture, and other agencies also have major statutory responsibilities in hazardous materials regulations and emergency response. The key challenges were: to develop data elements acceptable to all parties, to meet individual and collective needs, and to identify the forms as a Board product (a collective effort) in lieu of individual agency identifications.

The forms are labeled as Appendix A to this paper.

Part I is the Incident Report form. Information is needed by myriad state and local agencies immediately following a hazardous materials incident to assist organizations in implementing plans and initiating response to emergencies. Information initially reported includes: time of spill, injuries, product involved, meteorological information, and other information necessary to effect response decisions.

The Illinois Emergency Services and Disaster Agency had, in place, rules for the telephone reporting of hazardous materials incidents. Respondents were asked to continue the practice of initially notifying IESDA, via a 24-hour, toll-free number (1-800-782-7860), using the new forms.

Part II of the form is a single page, "Incident Evaluation" form. Evaluation is a key element in the Illinois Hazardous Materials Emergency Response Program. It is a post-response process. At the end of an incident, responders submit a report on the nature of the incident, its location, what chemical was involved as well as the role they played in the response process. They rate the quality and effectiveness of the response.

This type of evaluation generates valuable information that is of tremendous help to planners, policy makers, and responders, as they make important decisions that impact hazardous materials emergency response in the State of Illinois. It has specifically helped the state identify areas of weakness in the emergency response training and general preparedness for hazardous materials incidents. This has consequently led to significant changes and improvements.

For the scope of this paper, the evaluation process will be described in terms of its three component phases, namely: completion of the incident evaluation form; screening of forms and data input; and finally data analysis. At the end of such a description, some of the data that are typically obtained through such a program will be examined. That will be followed by a cursory look at how the data are analyzed. It is hoped that such an approach will provide a concise idea of hazardous materials incident reporting and data analysis in Illinois.

Incident Evaluation Form

This form is used uniformly across the state by all responders at the end of each incident. There are 40 items of information requested on the form. For a convenient discussion, these 40 items will be grouped into those referring to the responding organization, those relating to the incident, and finally a questionnaire of 19 items intended to rate the quality and effectiveness of response.

In filling out a form, the type of the responding organization must be indicated. It may either be private, municipal, county, or state. Since each organization plays a specific role during the response process, it is thus important that such a function be stated. Typically, an organization may be involved in one or more of the following: security/crowd control; spill mitigation; clean-up/recovery; medical assistance; firefighting; rescue; public information/warnings; and technical assistance/sampling. Moreover, a responder must specify his particular functions during the entire response process.

It is important for responders to describe an incident as thoroughly and accurately as possible. Therefore, the month, day, year, and time it occurred must be recorded. In addition, they must indicate: when their organization was notified; when they initiated response; when they arrived on the scene, and when all operations were terminated. An initial response can be made by providing information by phone or it can be an action on the scene of the incident. Responders are requested to indicate one of the two.

Many incidents result in injuries and/or deaths. These must be reported whenever they occur. The diversity of the areas where hazardous materials incidents occur almost parallels the ambiguity of the materials themselves. Thus, it is important for responders to also indicate where given incidents occur. These may include: a highway; a processing plant; pipeline; waterway; loading dock; oil field; rail; storage area; fixed facility; an agricultural facility; or any other area. Having indicated where an incident occurred, responders must characterize the chemical involved. This can be done by using the UN or NA number along with the chemical and/or trade name. Even when more than one chemical may be involved in an incident, the UN/NA number and name of each must be specified.

Hazardous materials exist either as solids, liquids, or gases; thus, responders must indicate a chemical's physical state. The type of container in which the chemical was present must also be indicated. Examples listed on the form include: truck; railroad car; drum pipeline; process pipe; and tank (aboveground and/or underground). Whenever possible, an estimate of the amount of materials at risk, the amount actually released into the environment, and the amount recovered must be made and reported in pounds, gallons, curies (for radioactive materials) or in cubic feet (for natural gas).

To keep track of costs of response activities in the state, responders indicate the total number of dollars and number of hours spent to handle an incident. As mentioned earlier, the incident evaluation process in Illinois includes a questionnaire. It is comprised of 19 questions at the end of the evaluation form. This questionnaire is intended to assess the preparedness of state and local units for hazardous materials incidents. The 19 questions relate to the following factors: timeliness of notification and response; appropriateness of state and local contingency plans; equipment availability; sufficiency of supplies; security and crowd control; staff training; intergovernmental cooperation; and general handling of incident.

These factors are rated on a scale of 1 to 8 as follows:

1 Does not apply
2 Strongly disagree
3 Disagree with reservation
4 Do not know
5 Neither agree nor disagree
6 Agree with reservation
7 Agree
8 Strongly agree

Screening and Data Input

Incident evaluation forms are sent to the Board, in care of the Illinois Department of Transportation. When the forms are received, they are first screened to make sure that relevant information is submitted. When a discrepancy or an omission is found in a report, a staff person contacts the responder or the individual who completed the report.

Each incident is assigned a number by the Illinois Emergency Services and Disaster Agency (IESDA). It is issued whenever an incident is reported and is usually indicated on top of the evaluation form. Quite frequently, more than one report is submitted for a single incident. Yet, in such cases, all of the reports are assigned the same incident number. This is due to the fact that multiple organizations often respond to an incident; thus, separate forms are needed. Data elements are separated in two categories: primary data elements (unique to the incident regardless of the reporting agency, i.e., information that doesn't change such as the chemical involved, number of injuries, location, etc.) and secondary data elements (information unique to the reporting organization).

After a report has been thoroughly screened, information is entered into a computer data base which is closely monitored. Such information is periodically retrieved, updated, and presented to the Board as a status report on hazardous materials incidents in Illinois. Moreover, it constitutes a significant part of the Board's annual report to the Governor.

Since the inception of the incident reporting and evaluation program in Illinois, various kinds of data have been generated regarding hazardous materials incidents. Examples of such data are: information about responding organizations; number of incidents (at state and county levels); origin of incidents; number of incidents by container/mode of transportation; chemicals most often involved in accidents; amounts of hazardous materials at risk, released and recovered in an incident; injuries and deaths; and the cost of clean up operations.

In the sections that follow in this paper an attempt will be made to identify some ways in which data obtained from the incident evaluation form are analyzed. The data referred to were obtained from the 1988 incident reports presented to the Board. These were also published in the Board's 1989 report.

1. Number and Types of Responding Organizations

    The Board received 4,221 forms in 1988 in which 2,113 hazardous materials incidents were reported. Most of the forms (about 78 percent) came from Illinois state government organizations. The rest came from the following types of organizations: municipal (14 percent); private (7 percent); county (1 percent); other government (0.7 percent); a small percentage (0.2 percent) of organizations were not specified.

    The Illinois hazardous materials incident notification and evaluation program is standardized. Organizations that deal with hazardous materials in the state are expected to participate in it. Therefore, it is important that figures pertaining to those response organizations be presented in a report. Most of all, such an analysis helps the Board assess the level of participation, by organizations, in the program.

    Most of the responders who submitted forms to the Board in 1988 were from emergency management (about 52 percent). Illinois EPA responders submitted about 25 percent of the forms; fire departments 13 percent; spillers or their representatives, about 6 percent; and law enforcement, about 3 percent. The remaining few came from hazardous materials contractors, emergency medical personnel, and others not specified. This type of breakdown of data reflects the diversity of roles or activities typically involved in a hazardous materials incident response.

2. Number of Incidents by Location and Container/Mode of Transportation

    Data on the sites in which hazardous materials incidents commonly occur can also be obtained. For instance, in analyzing the 1988 data, it was observed that about 41 percent of the reported 2,113 incidents occurred at fixed facilities. Highway-related incidents reported accounted for about 15 percent whereas about 4 percent of those reported were railroad related.

    Such an analysis enables policy makers to prioritize emergency response and enforcement plans.

3. Hazardous Materials Involved in Most Incidents

    Knowledge of the types of hazardous materials involved in incidents across the state is very important. As such, incident reports are analyzed to present information on materials that are most frequently involved in accidents. Because of the possible inaccuracies in volume estimations by responders, only the number of forms was analyzed in which respective chemicals are reported. Gasoline was the product most frequently involved in incidents (reported in 398 forms) in 1988. Next were combustible liquids (272), and petroleum crude oil (131). Some others included: PCB (86); oil, petroleum (70); sulfuric acid (42); anhydrous ammonia (39); natural gas (39); hazardous substance N.O.S. (38); and hydrochloric acid (33).

4. Injuries/Deaths; Man-hours; and Replacement Costs

   The severity of hazardous materials incidents has always been a heated topic of discussion. Therefore, the number of deaths, and the number of injuries incurred in hazardous materials accidents are recorded and analyzed. Figures are presented at both state and county levels. For instance, in 1988, 12 deaths and 832 injuries occurred in hazardous materials accidents throughout the state. A breakdown of these figures will give an indication of the counties in which deaths and injuries occurred. It is important to note that these numbers do not represent deaths and injuries from strictly the chemical(s); rather due to the incident as a whole.

   The severity of an incident can also be interpreted in terms of the number of man-hours and dollars spent on cleanup operations. We now have a much clearer idea of the magnitude of state and local response. Further, the program can present information on the environmental impact of hazardous materials incidents. This is done by presenting data on estimated amounts of materials at risk, released into the environment, and estimated amounts recovered. One can estimate the amount of materials unrecovered from the environment by subtracting the amount recovered from the amount released.

   As mentioned earlier, the incident evaluation program used in Illinois allows responders to rate various aspects of the state's preparedness for hazardous materials incidents and also the cooperation between various organizations involved in hazardous materials emergency response. In analyzing data obtained in 1988, for example, most (about 85 percent) of the responders felt that the response of state agencies to incidents was timely; 86 percent indicated that local government's response was timely; and 63 percent felt that they received a timely notification for response. However, data in 1988 also indicated that there was need for more inter-agency cooperation.

   From the above, one can see that the information obtained and analyzed about hazardous materials incidents in Illinois is comprehensive. Analysis of the type of data obtained presents a good indication of the location, nature, and frequency of hazardous materials incidents in the state. Moreover, it spotlights the strengths and weaknesses in the state's emergency response program.

Recommendations for the Illinois Program

The hazardous materials incident reporting and evaluation program has been viewed as a success in Illinois. Much more is now known about the frequency and severity of hazardous materials incidents in our state. This information and analysis assists state and local planners. Several improvements are recommended specifically for the program in Illinois:

- The percentage of participation by units of local government needs to be increased. More direct and frequent contact with individual police and fire departments may be needed. The private sector should also increase its participation.

- The data base should be available for direct access by state and local agencies. Additional computer equipment and software development may be needed to facilitate increased access.

- The Illinois program should be specifically funded by the General Assembly to ensure that needed upgrades can be made and the program promoted within the state.

The evaluation program integrates with the other comprehensive state programs. The process has brought state agencies and local organizations closer together. Standardized reporting greatly assists in a comprehensive planning effort. Government and industry now work more closely through a public/private partnership. The other innovative programs as discussed earlier in this paper clearly demonstrate how government and the private sector can work effectively together for the mutual good, without unreasonable burdens or costs on industry.

Where We Go From Here: Recommending a National Strategy

Despite the growth in computerized data bases, there are significant challenges that must be met to increase the access of data. The following recommendations are offered as opinions of the authors to develop a more comprehensive national strategy:

1. Incident reporting systems and data bases maintained in other states and on the federal level should be the topic of a national level study. There is a need for sharing information to assist states in developing new systems or upgrading current ones. More aggressive promotion is needed of existing federal hazardous materials incident data bases. State agencies and local governmental units should be encouraged to utilize data currently available. Unfortunately, those resources are not widely understood. Federal funding should be appropriated for this effort.

2. A public/private partnership is needed to develop an integrated national level program that could benefit both government and industry. A federally funded task force should be appointed for the

purpose of identifying state and local needs, reviewing current programs, and recommending an integrated approach to maintaining and accessing a hazardous materials incident data base. There are several good programs in place, like the Hazardous Materials Information System; however, a more active approach is needed maximize the use of systems. The study should identify the basic configurations (hardware, software, modem, etc.) needed to fully utilize these vital services. The study should explore the potential need for nationwide computerization program for response organizations, the cost of such a system, funding mechanisms, and any services that should be developed in the future. Key state and local organizations that have a major role in hazardous materials planning, training, and response should be contacted to identify their computer capabilities. It is safe to assume that most response organizations do not have computers; over 80 percent of fire departments are volunteer in the United States and generally have small, inadequate budgets.

3. Tie in with the National Fire Information Council to identify the relationship between existing state reporting systems and the new hazardous materials incident report program, now available to the fire services nationally. This program was initiated in January 1990 and features a single page report form.

Much has been accomplished in Illinois, and on the national level. However, additional needs must be met. A state-by-state or city-by-city approach is not necessarily the best one to follow. In our opinion, a much more proactive, public/private partnership is needed to enhance the planning, training, and response needs of our communities. A coordinated program could prove to be more cost effective. Federal level leadership in this initiative may be the best approach for a unified national system.

Canadian Database Development as a Support Tool to
Transport Risk Assessment

Donald A. Learning [*]

Abstract

The Risk Management Branch of the Transport Dangerous Goods Directorate supports a multi-modal safety program for dangerous goods transportation through the measurement, assessment and management of risk.

To achieve this, the Branch has developed a series of databases which provide information on commodity flow, accidents and compliance activities, and is currently completing modelling work to estimate the economic and social costs of dangerous goods accidents. Together, this information will allow the Directorate to better define risk issues and better assess alternative risk management techniques.

This paper provides details on the databases the Risk Management Branch has available to assess accident trends and exposure to dangerous goods, and introduces a costing model under development to further enhance risk management capabilities.

[*] Transport Canada, 344 Slater Street, 14th Floor, Ottawa, Ontario, Canada K1A 0N5

Introduction

In the context of public service and governmental performance, the goal of the Transport Dangerous Goods Program is to reduce risk in the handling, offering for transport, and transport of dangerous goods. To do this it is necessary to apply risk management techniques in a regulatory framework targetted towards a highly diverse and competitive sector of the Canadian transportation system.

Risk has two components; probability and consequence. By reducing the frequency of dangerous goods accidents, risk is reduced. Likewise, by reducing the consequences (losses) of non-preventable accidents, risk is reduced. As a governmental body, the Transport Dangerous Goods Directorate is challenged to effect changes in accident frequency and consequences in a cost effective and economically efficient manner.

In meeting this challenge, the Dangerous Goods Directorate collects and analyses data on commodity flows, DG accidents, and on incident costs related to these accidents. This information, is then assessed as a measure of the current risk associated with the transportation of dangerous goods. Risk management actions, such as regulations, can then be designed to effect the greatest improvement in safety (reduction in risk) with the smallest societal cost. That is, to manage risk in a way that is both cost effective (ie. benefits exceeding costs) and economically efficient ie. the least cost alternative to effecting the desired results.

In Canada, the reporting requirement for dangerous goods accidents, Section 9.14 of the Transportation of Dangerous Goods Regulations, came into effect on July 1, 1985. This requirement obliges the employer of the person in charge of a dangerous goods load at the time of an incident to report when:

    1.   there is a loss of dangerous goods in defined levels that represent a danger to

health, life, property or the environment;

2. when a bulk means of containment is damaged;

3. when radioactive materials are involved in a transportation accident;

4. when there is an unintentional fire or explosion involving dangerous goods;

5. when a person has been injured or killed in an accident in which there has been a release of dangerous goods;

6. when damage to the integrity of a pressurized means of containment is discovered;

7. when there is a suspicion that a container hs suffered damage to its integrity from impact, stress or fatigue; or

8. when all or part of a consignment of explosives or radioactive material has been misplaced, lost or stolen.

Important as it is, accident information, particularly accident frequency information, is of little use without some measure of accident exposure. That is, a means to calculate accident rates and by comparison evaluate if accident rates are increasing, decreasing or remaining static. A second series of databases is thus maintained to provide one such exposure measure — the tonnage of dangerous goods shipped, or, "commodity flow."

Unlike the accident data, commodity flow information is not collected through a legal reporting requirement. Rather, road flow is estimated in cooperation with Statistics Canada's for-hire trucking survey, marine flow is estimated from shipment permits issued by 6 of Canada's major ports, and rail flow is estimated from data provided by Canada's two national railways.

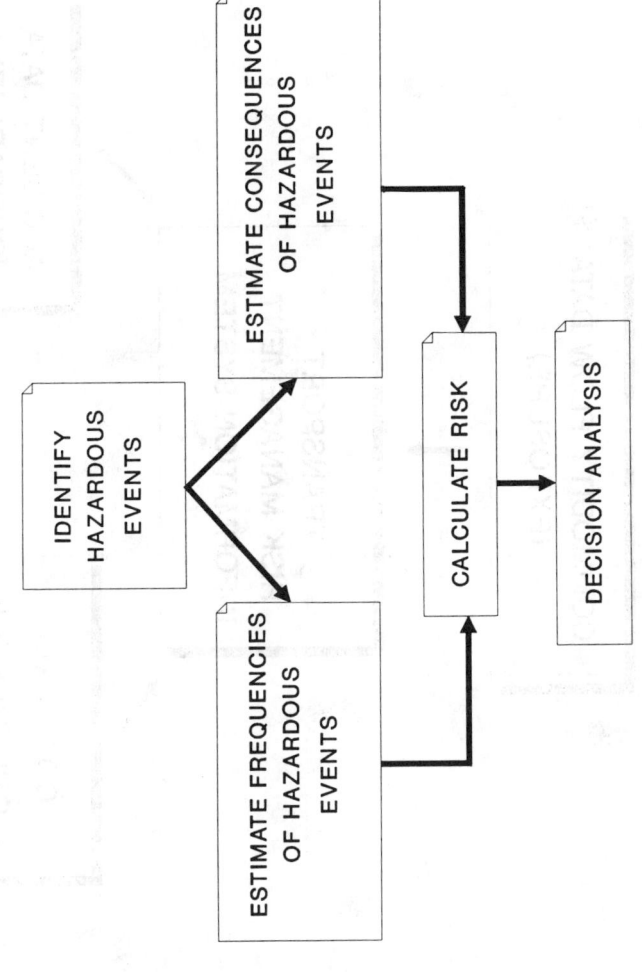

Figure 1

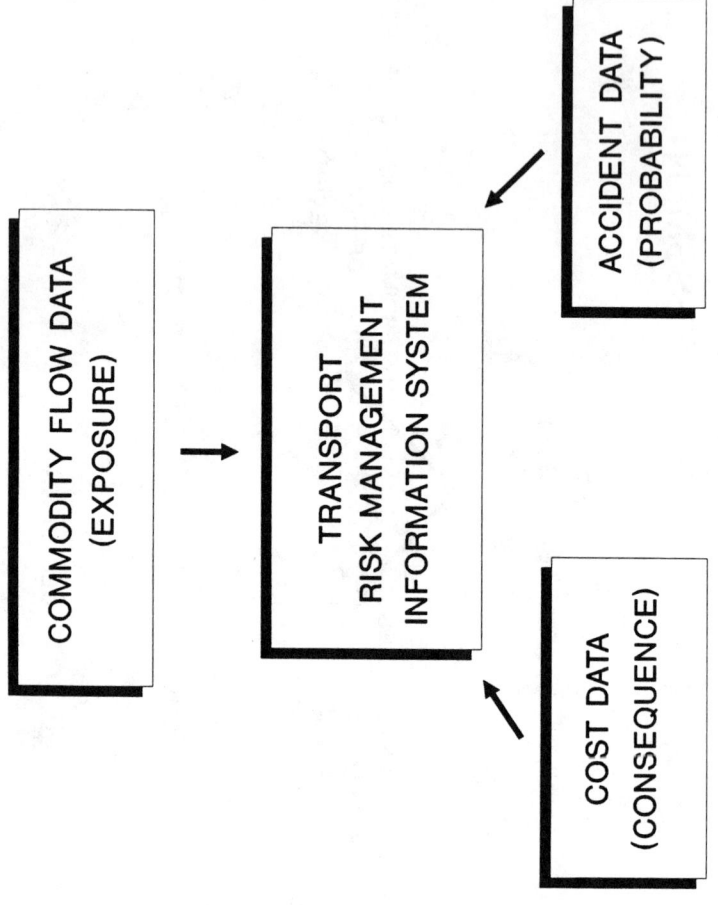

Figure 2

Risk Analysis

The ultimate goal of any risk management information system (RMIS) is to provide a tool for reducing uncertainty in decision-making - ensuring that decision-making is not done in ignorance.

The Dangerous Goods Directorate's in-house analyses have historically responded to requests for information researched on an ad hoc basis plus the preparation of regulatory impact analysis statements (RIAS). These tasks, while necessary for program maintenance, are insufficient to clarify risk issues and initiate risk reduction activities.

What is necessary is an integrated risk management system capable of gathering, collating and analysing a wide range of data, from across the country, on a consistent and regular basis.

Risk management must respond to the basic objectives of the organization. For industry these include:

- profitability
- ensuring uninterrupted service despite accidental losses
- maintaining stable earnings
- achieving growth in sales or market share
- complying with legal requirements, and
- fulfilling humanitarian concerns.

The risk management objectives of a national government, however, are much broader.

With industry cooperation, and regulations and enforcement as tools, the Directorate is expanding upon its current flow and accidents databases towards the implementation of an overall Risk Management Information System. This system addresses all modes of transport and is sensitive to risk transference effects between them. It covers all of Canada's ten provinces and two territories, and is being designed to impute information not currently available to us. In addition to this, the design allows for risk reduction options to be modelled and tested for effectiveness so as to provide comfort that those options chosen operate as intended.

Risk Identification

The first step in the risk management process is to identify all possible exposures to loss. The organization or government must know, in effect, "what can bite them."

A risk management information system based on past experience identifies areas where losses may occur, allows for the examination of alternative actions for reducing these losses, and allows monitoring of the results of the alternative risk reduction actions that are implemented.

To carry out this iterative process, the components of risk must be identified and analysed. Data on property damage, net income losses, liability losses, and persons affected must be broken down and described as exposure types.

The databases put into place by the Risk Management Branch are intended to do this. The "accident" database contains information on evacuation, types of contamination, whether the area involved is rural, urban or commercial, and a host of other details related to a particular incident.

These systems enable us to collate exposure information on dangerous goods shipments and estimate accident probabilities from dangerous goods accident report data. Accident cost follow-up and modelling provide the third element – information on accident consequences. Together, commodity flow, accident frequency and cost information provide generalized frequency and severity measures (probability and consequence) for risk assessment.

Risk assessment is as much an art as it is a science and involves the modelling of losses under varying types and levels of exposure. That is, given a certain level of goods transport, the accident frequency represents a probability distribution of a loss exposure. The losses related to these types of occurrences represent the

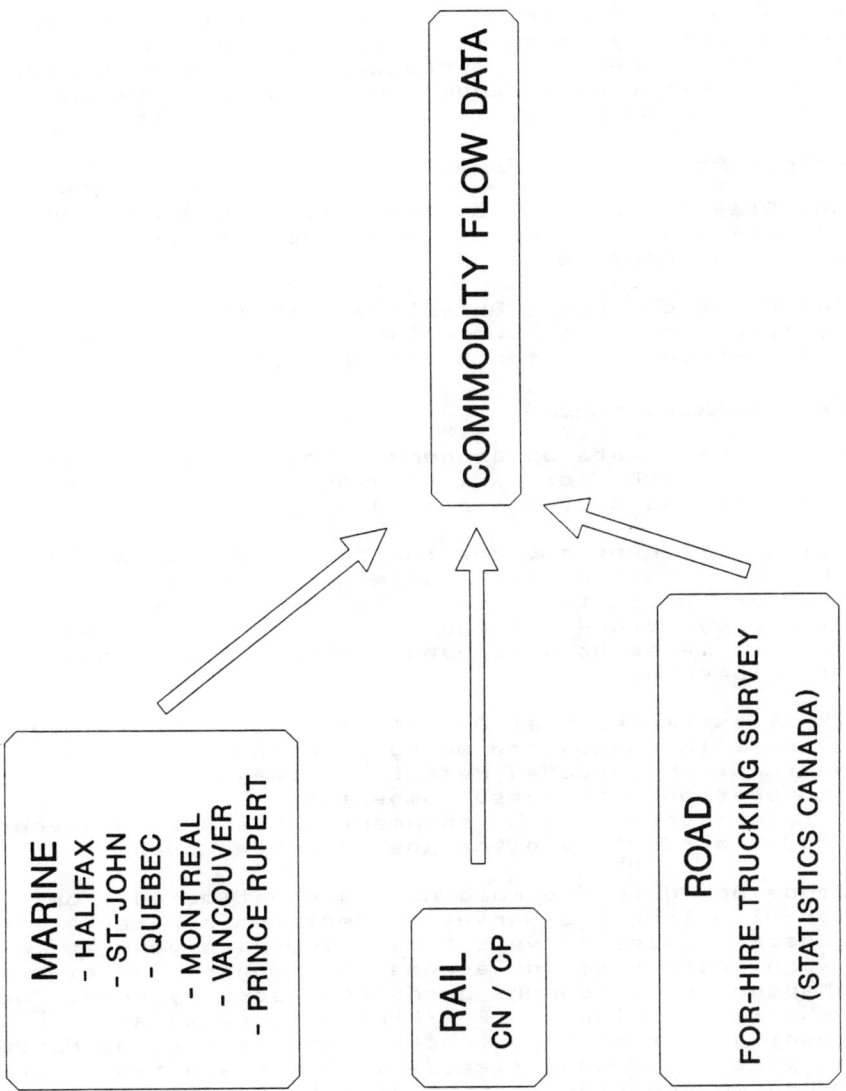

Figure 3

consequences for which risk management solutions are sought. In simplistic terms, management can be focussed either upon a reduction in the probability of occurrence or a reduction in the cost of the outcomes — or both.

Databases for Risk Management

Any Risk Information System relies upon up-to-date information representative of the activities that are to be managed.

The Dangerous Goods Directorate fulfills this need by maintaining its own databases and by surveying information collected by other agencies.

Commodity Flow Data

Directorate data on dangerous goods flows go back as far as 1981 for rail transport and as far as 1984 for the marine and road modes.

Using different sources for each mode, a matrix has been constructed which estimates total goods transported in the country by year. From these data projections are made as to how much of what product is being moved and the origin-destination of shipments.

It is estimated that 80% of dangerous goods moved by rail in Canada are moved by either Canadian National or Canadian Pacific Railways. In cooperation with these companies, origin-destination of shipments, class of dangerous goods, product, and tonnage is broken down.

Goods moved in the road mode are estimated from for-Hire Trucking Surveys undertaken by Statistics Canada. These surveys take place each year at weigh scale stations across the country and give a shapshot of dangerous goods transport by road. The vehicles sampled are For-Hire only, however, meaning that private trucking (transports operated as part of company fleets) is not evaluated. Since For-hire trucking is estimated to be 40% of road traffic, the sample is adjusted to reflect what total road traffic would be.

Marine dangerous goods traffic was estimated until 1987 by manually inputting and counting dangerous goods shipping permits provided by the port authorities of 6 of Canada's major ports. These ports, Halifax, St. John, N.B., Quebec, Montreal, Prince Rupert, and Vancouver have now begun to provide to us data tapes of this information.

Dangerous waste movements are being estimated from data collected and provided by provincial governments. Currently information is being supplied by Quebec and Ontario; two provinces in which a large percentage of Canada's dangerous waste is transported.

The data collected on commodity flow require considerable work on the part of Directorate staff to convert them into useable origin-destination format with consistent weight and volume measures. As a result, the most recent data available on flow lags up to 5 years.

The apparent seriousness of this timing problem is offset somewhat by the fact that commodity flows are increasing at only about 3% per year. When accident rates are generated using these older flow data, the rates can be considered very conservative (near-current accident frequency and 3-5 year old tonnages).

Accident Databases

As noted earlier, the Dangerous Goods Accident Information System stores information gathered from reports which must be filed with the Directorate within 30 days of an accident.

The report form allows those responsible for dangerous goods loads at the time of an accident to advise the Directorate of their assessment of the circumstances, suspected causes and outcomes surrounding an incident.

Accident information is used to point to safety weaknesses. Comprehensive, directed research of accident trends must be effected, however, before setting standards and promulgating regulations.

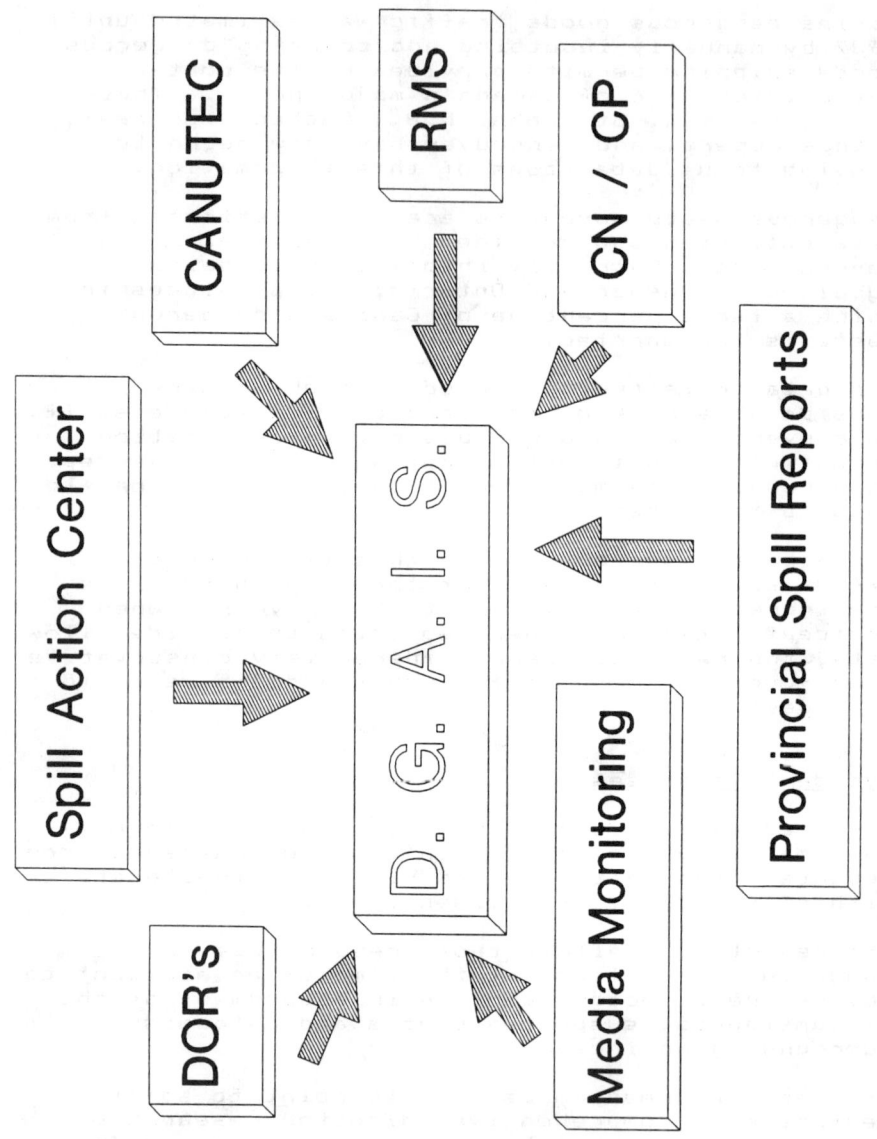

The measurement of accident consequence is equally important for risk management. For this reason, the accidents database contains many fields dedicated to describing accident outcomes - amount of product lost, amount of product carried, release locations, deaths, injuries, evacuations, emergency responders on scene, etc.

Follow-up of accident reports is undertaken to collect information on property damages. Last year a staff of one full time analyst and a clerk processed over 1000 reports. This volume obviously places considerable constraints on the time available to analyse reports and to add details to the database.

To solve this problem a project to estimate costs, from accident descriptors, through a modelling effort, has begun.

Consequence Information - Costing

Cost information is crucial for estimating the consequence element of risk.

Early in 1989, the task of cost information collection was begun along two fronts. This comprised the in-depth study of the costs of 70 major dangerous goods accidents as well as the follow up of road mode accident reports. These efforts promise to provide a good measure of the economic losses associated with dangerous goods accidents.

The information is being incorporated into a forecasting model which can use accident descriptors collected from accident reports to estimate costs for dangerous goods accidents.

At the present time the fields on the accident database tell us how much of what types of product are involved in accidents, how much is released, the emergency response that resulted and the types of contamination that occurred. This and other information coded on the database can be compared, among other things, to unit costs for emergency response personnel and equipment, evacuation costs

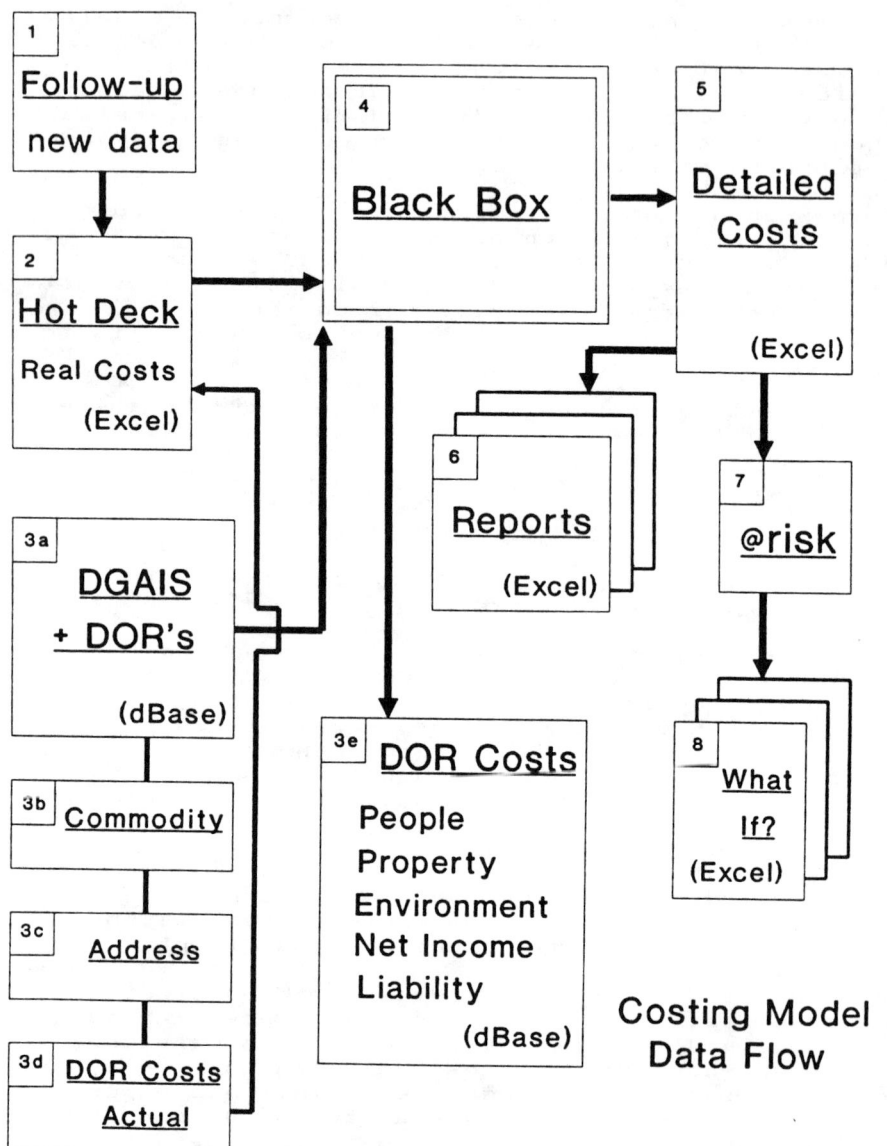

Figure 5

by region, type of property and time, and transport equipment damage, shipment value, and loss of shipper revenue.

For modelling and planning purposes, "potential accident" costs, can be estimated from historical data generated from accident analyses. During risk assessment, the effects of catastrophic but rarely occurring events can be estimated in the absence of historic data through simulation.

Summary

The Directorate meets its information needs by constructing specialized datasets from many different information sources.

Using data available outside of the department, a profile of dangerous goods commodity flow is synthesized. Where useful data cannot be imported (accidents, for instance) a legally mandated reporting requirement has been implemented and centrally managed within the Directorate.

Cost information is being investigated through in-depth investigation of the costs associated with significant accidents over the past several years. Item costs including the type of product involved, level of emergency response, mode, and other accident particulars are being modelled in concert with researched costs of accidents such that for each accident recorded by the Directorate, an accurate total cost estimate for future accidents, based on experience, can be generated.

As the model is implemented and datasets improved, the Directorate's risk management information system will ensure that dangerous goods transportation risk in Canada is managed in a logical, realistic, and justifiable way.

Its application in risk management will ensure increased safety for the Canadian public.

Summary of Risk Databases

(a) Accident database: The Dangerous Goods Accident Information System is a database of accident reports collected under law since 1985. The data base has over 2200 records, each of which contain 74 fields to describe, inter alia, accident location, commodities involved, events leading up to the accident and the consequences in terms of injury, death, evacuation and spillage.

(b) Commodity flow database: The DGFlow database contains information collected from Canada's two major railways, DG shipment permit information from 6 of Canada's major ports and road transport data from Statistics Canada's for-hire truck traffic survey. Air flow has not yet been accurately surveyed but is estimated as a percentage of total flow in other modes.

(c) Cost database: The cost database captures accident cost (consequence) information gathered through followup of accidents reported under Section 9.14. This sampling of acute accident costs supports modelling work and when combined with modelled costs will ensures that for each incident assessed by the Directorate can have a consequence or cost, component.

RISK ASSESSMENT/ROUTING

MANAGING RISKS OF HAZARDOUS MATERIALS TRANSPORTATION IN SANTA BARBARA COUNTY

Douglas K. Anthony, AICP and John Peirson, Jr.[1]

This paper presents some potential building blocks which local jurisdictions can pursue to manage the risks associated with hazardous materials transportation (HMT). The recent experience of Santa Barbara County, California, serves as our reference point. Federal research programs and regulations have undoubtly gone far in reducing HMT risk. Among other things, they provide for safer containment and better identification of hazardous materials in transit. There are several aspects of risk management, however, where local jurisdictions can be quite effective and efficient. Some local jurisdictions, for example, have implemented routing restrictions to minimize public exposure to potential risks. Additionally, several jurisdictions have developed more sophisticated capabilities for responding to HMT incidents.

These two examples perhaps preface the importance of a responsible local role that custom tailors risk management strategies to fit local conditions. The local role, to be sure, is not a new concept *per se*. It is in an early developmental stage of a dynamically evolving process, however, being driven by a vigilant public that demands a safe, clean, and healthy environment.

This, at least, is the perception of the local role in Santa Barbara County, a region where HMT risks have long ranked high among public health and safety concerns. Public concern and awareness stem, in part, from three factors: (1) a relatively high and still increasing volume of HMT compared to the county's small population of 348,400; (2) the acutely hazardous properties of particular materials transported, such as rocket propellant and anhydrous ammonia; and (3) a general

---

1. Douglas K. Anthony, Planner, Resource Management Department, County of Santa Barbara, 1226 Anacapa Street, Santa Barbara, CA 93101; John Peirson, Jr., Director of Environmental Health and Safety Practice, Arthur D. Little, Inc., Creekside Plaza, Suite 220, 5290 Overpass Road, Santa Barbara, CA 93111-2050.

belief that transportation risks could and should be better managed. As a result, the county has been developing risk management strategies to address both general factors and more specific, localized factors of risk. These efforts have provided a learning experience which may offer some interesting insights to others. As we look forward to learning from the experiences of other participants in this national conference, we also wholeheartedly invite critique of our own approaches to managing HMT risks.

LOCAL SETTING

Santa Barbara County is situated in the southwest portion of California's central coast region, as indicated in Figure 1. The county exhibits a predominantly rural character with approximately 39.5 percent of land dedicated as national forest and another large portion designated for agricultural use.

Regarding major state freeways, Santa Barbara County lies substantially west of Interstate 5 -- the primary north-south artery in the state -- and is separated from this artery by a coastal transverse mountain range. U.S. Highway 101, which provides the major north-south link for the central coastal region, dissects the county as shown in Figure 1.

These same characteristics apply to rail linkages. The primary north-south rail line in the state runs up the middle of the state, while the coastal rail line dissects Santa Barbara County.

Three major highway routes are available for easterly shipments of hazardous materials. These routes, illustrated in Figure 1, are State Route 46 to the north, State Route 166, and State Route 126. Regarding rail, shipments bound for easterly directions are first routed to rail yards in Los Angeles, and then rerouted to easterly destinations.

Highways in Santa Barbara County experience the typical types of HMT such as gasoline, aviation fuel, chlorine, and household and commercial hazardous wastes. Like several cities and counties across the nation, however, highways in Santa Barbara County also experience some atypical volumes or types of HMT as summarized below.

* The Casmalia Resource Hazardous Waste Facility, a Class I disposal site which, until recently, received truck shipments of hazardous wastes from numerous California counties and out-of-state generators. The facility operator recently applied

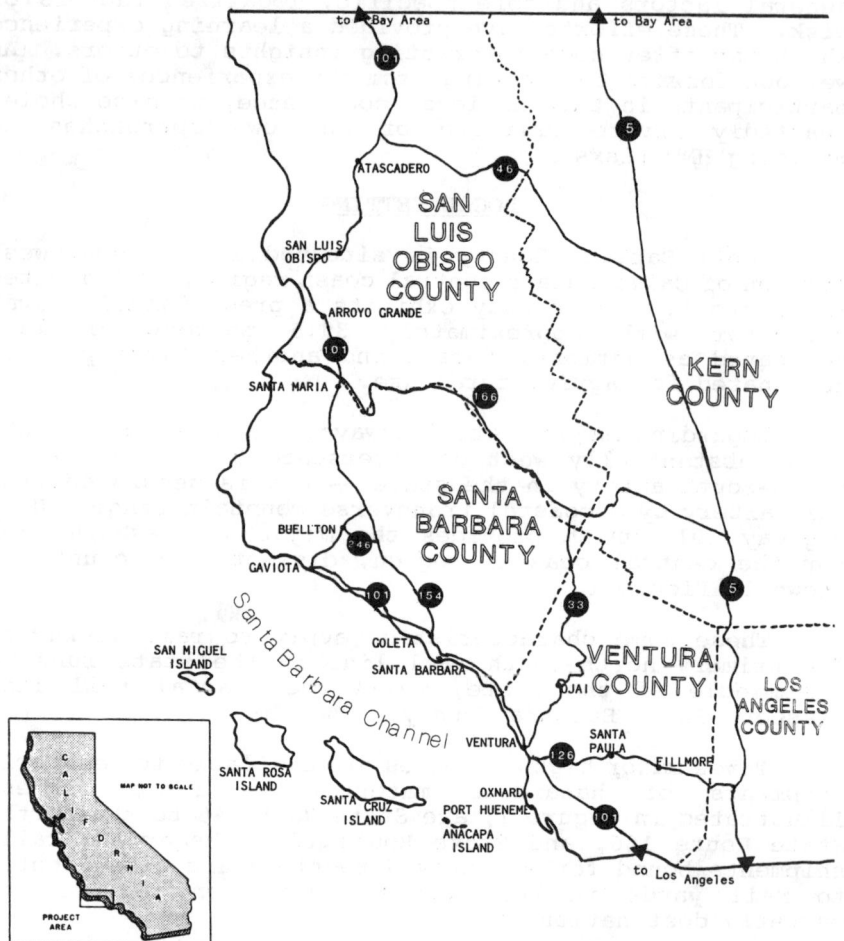

**FIGURE 1
COUNTY OF SANTA BARBARA**

for new permits to develop new landfills that would expand the number and types of waste that the facility can handle if approved.

* One of the only two major launch bases for space orbits nationwide, Vandenberg Air Force Base also serves the nation as a major missile test range. To support these operations, the base receives truck deliveries of hazardous hypergolic propellants such as nitrogen tetroxide (a Class A poison), hydrazine, and Aerozine-50.

* Increased emphasis on developing oil and gas reserves offshore, resulting in more daily truck trips to ship such hazardous materials as sulfur, liquefied petroleum gases, and heavier gas liquids such as gasoline to destinations outside the county.

* Recently developed measures to control emissions of nitrogen oxides into the air originally sought to require the Best Available Control Technology (BACT). After further analysis, however, regulators discovered that BACT required use of anhydrous ammonia in notable quantities, which meant several annual shipments of this extremely hazardous substance on our highways.

The county's exposure to hazardous materials transportation has continued to increase our awareness not only of associated risks, but also of options to reduce these risks. It has been and continues to be a learning process, beginning in 1985 with a routing restriction placed on hazardous waste shipments to avoid potential releases near a critical water supply. Subsequently, we have expanded our efforts towards a more comprehensive approach to risk management, which we summarize below.

PHASE I - FOUNDATION:
ESTIMATING RISK AND RISK REDUCTION

Developing a local risk management program begins with some very important choices: which approach, process and method for estimating risk and risk reduction best fit local needs and circumstances. Our experience suggests that local and regional risk assessments, if designed and executed properly, can go far in identifying specific sources of risk and appropriate risk-reducing measures. The text below summarizes what we believe are some basic building blocks to effectively manage local transportation risks.

(a) Select a manageable approach.

To develop a local risk management program, we began by taking a piecemeal approach to risk assessment while attempting not to lose sight of the cumulative implications of HMT in our region. Although funding constraints gave us little choice when selecting the piecemeal approach, we find that it has several advantages. It involves preparation of several risk assessments, each of which is limited in scope to studying a particular type or group of hazardous materials. Each risk assessment is built into a single computerized model for the region. This approach helps us to keep our heads above water while avoiding a "one-fell-swoop" approach to HMT risk management which can be overly broad and general, leading to confusion, over-simplification, errors in risk estimation, and ineffective management of risks.

One of our risk assessments, for example, focused on the gas liquids (propane, butane, and heavier gas liquids) that are produced from oil and gas fields in and offshore the county. These materials are classified as hazardous by virtue of their flammable and potentially explosive characteristics. By limiting our scope thusly, we were able to focus on specific characteristics of gas liquids, on the potential routes that would be used to ship these materials, on the specific operations of gas liquids' carriers, and so forth.

Another risk assessment currently underway focuses on the transportation of hazardous wastes, which carry notably different types of risk compared to gas liquids. Consequently, not all risk reducing measures that are found to be appropriate for transportation of gas liquids may be suitable for hazardous wastes. While avoiding heavily populated areas weighed substantially in routing considerations for shipments of gas liquids, avoiding routes adjacent to major reservoirs or groundwater recharge zones are important when considering routes for hazardous waste shipments. In this case, what may be the best choice of routes for one type of hazardous material may not provide the best choice for other types of hazardous materials.

The computerized model, discussed further in item (h) below, was developed as part of the first risk assessment. It contains several data banks and analytical components, including route-specific accident rates, conditional spill probabilities for different types of containers, and estimated consequences if an accidental release occurs. Subsequent risk assessments are then built into the same model. The overall model, including the data, can be revised as required to account for changes in the transport system over time.

(b) *Ensure effectiveness and efficiency through geographic and situational specificity.*

The ability to address the specific geographic and situational characteristics of HMT as it occurs in a particular region is one of the most important contributions that local jurisdictions can make towards risk reduction. Therefore, the approach, process, and method employed to estimate risk and risk reduction should focus on such characteristics.

Particularly, structure the risk assessment to be mode-specific and route-specific, reflecting the actual conditions that characterize the regional HMT system as much as possible. Avoid using generic risk assessments unless properly adapted to fit the local situation. The generic risk assessment can provide useful insight into those risk factors that are independent of geographic and situational factors. For risk factors that are dependent on geographic and situational specificity (e.g., actual routes, number of shipments, number of rail cars shipped on one train, etc.), however, the generic risk assessment can easily mislead, resulting in risk managing strategies that do little or, worse yet, increase risk.

(c) *Ensure effectiveness, efficiency, and creativity by bringing shippers, carriers, and governmental agencies into the risk management process early.*

We have found that working with an advisory group contributes immeasurably towards achieving geographic and situational specificity in our program. The group comprises shippers, carriers, enforcement agencies (such as the California Highway Patrol), emergency responders, transportation agencies, and neighboring jurisdictions.

Additionally, the broad expertise that this group brings to the process of risk management provides several advantages. This broad range of experience, for example, adds considerably to the level of creativity that we are able to bring to the process. Additionally, committee meetings provide an excellent forum for stakeholders to express concerns openly, to clarify potential misconceptions, to develop professional respect for one another, and simply to share responsibility in risk management. The diversity of representation on the committee brings timely scrutiny of the risk management program as it develops, encourages the development of options, and promotes credibility and good faith. Lastly, the committee facilitates implementation and subsequent evaluation of the program.

(d) *Avoid one-dimensional estimates of risk that may later distort conclusions and lead to ineffective risk management.*

In developing quantitative risk assessments, various types of risk measures can be used together to present a clearer, overall picture of risk. One example is risk profiles which demonstrate societal risk levels overall; that is, the annual chance of an activity to cause a single injury or fatality, the annual chance to cause ten injuries or fatalities, and so on. This measure of risk is comprehensive and can include risks distributed over a range of physical locations.

In addition, it is often desirable and prudent to know at what level of risk a given individual may be exposed. This type of risk measure, known as individual risk, allows for ready comparison with background risks. Comparison with background risks should not be used to indicate whether the risk in question is acceptable or not; rather, it helps to put the risks in perspective.

More importantly, however, the use of individual risk provides a mechanism to measure and evaluate the issue of equitable distribution of risk. As mentioned above, a route identified through risk analysis to exhibit the lowest overall societal risk for a given material may have the highest individual risk, especially if many other hazardous materials are transported on the same route. Such a result would indicate that the individuals along this route are being exposed to levels of risk higher than other people within the transportation assessment area. Therefore, in any quantitative risk assessment that is used for assessing specific transportation routes, evaluation of both societal and individual risk levels is critical.

(e) *Go beyond simple comparison of alternative routes to identify sources of risk on specific routes and to measure the effectiveness of mitigation.*

Regional and local risk analyses often have been limited to a comparative assessment of alternative highway routes. Upon determining which route(s) offers the least or most risk to the public, appropriate routing restrictions are implemented. Although restricting HMT to routes that offer the least societal risk, it does not necessarily contribute towards the reduction of reduce individual risk.

This point was made very clear to us by a concerned public that lived and commuted along State Route 166 when the state classified this highway as an Explosives A route. The designation was based on a risk assessment which the U.S. Air Force conducted to determine the best routing scheme to transport extremely hazardous rocket propellant to its Vanderberg base on the west coast.

From a societal risk perspective, the risk assessment and subsequent routing restriction reduced risk by rerouting shipments away from more heavily populated areas (previously shipments were routed through Los Angeles). Considering individual risk, however, the risk assessment was completely silent; the subsequent designation, at best, shifted the individual risk without considering potential route-specific hazards and remedies. A subsequent risk analysis for gas liquids transportation on the same route revealed several effective measures that could be taken to eliminate or reduce clearly identifiable hazards along this route.

Risk assessments should go beyond mere comparison of alternative routes to determine route-specific hazards and remedies. This step is most important. Routes that avoid heavily populated areas tend to be routes that are not well-designed, not well-maintained, and not well-patrolled, and may also be hard to reach with emergency services. It also may be that, once route-specific remedies are identified and implemented, the comparison of alternative routes may bear different results.

(f) Evaluate Risk Reducing Measures.

A key part in developing an effective HMT risk management plan is the evaluation of risk reducing measures. This evaluation is necessary for two main reasons. First, it allows one to estimate the effectiveness of each measure in reducing societal and/or individual risk, and second, it allows one the opportunity to assess the technical and economic feasibility of each measure. Depending on the type of risk reducing measure, this evaluation can be either quantitative or qualitative.

Risk reducing measures can be broken down into three major categories: procedural measures, technological measures, and route-specific measures. Procedural measures would include areas such as driver training and hiring programs, vehicle inspection programs, drug and alcohol testing programs, etc. These types of measures can generally be classified as management systems, and need to be implemented as part of a risk management program.

The expected effectiveness of some of these measures can be estimated based upon information collected by transportation companies that have instituted similar programs. Figure 2 shows the results of a risk reducing measures analysis for truck transportation of propane to Los Angeles from Santa Barbara County. The results of this analysis showed that the use of improved driver training and hiring practices had the greatest potential for reducing risk. By evaluating data from companies that had instituted such programs, it was found that these types of programs reduced both the accident frequency and well as the likelihood of a release given an accident.

Technological measures are those associated with changes in equipment design that reduce risk such as using MC-331 trucks to transport natural gas liquids. Also included in this category are technological systems that can be used to improve the safe operation of the transportation vehicle. Examples of such systems would include vehicle monitoring systems (VMS) and vehicle tracking systems (VTS) for trucks. Due to the limited use of these systems today, it is very difficult to quantify the effect they might have on reducing transportation risk.

Another example would be the use of compartmentalized trucks. The effect of this measure on risk is shown in Figure 2. Use of this measure could reduce the consequences of a release given an accident. Figure 2 also shows how the use of smaller trucks could effect the overall risk. As can be seen from the figure, this measure could actually increase risk as a result of the increased number of truck trips required.

These two examples emphasize the need to include risk reducing measures in any HMT risk assessment. While both these measures could reduce the volume of material released in an accident, the use of compartmentalized trucks does not require an increase in the number of truck trips to move a fixed volume of material.

Route-specific measures would include things such as improved bridge construction, the addition of truck climbing lanes on steep long grades, improved intersection construction, etc. Also the use of radar for speed enforcement on local roads, while difficult to quantify, is considered a practical measure that has the potential of reducing both the accident rate as well as the likelihood of a spill given an accident. Route-specific measures typically result in reducing the accident rate, and are not focused sloely on hazardous materials carriers. For much of the effected population, these types of measures are viewed as critical in reducing overall vehicle accident rates on local roads.

RISKS IN SANTA BARBARA COUNTY

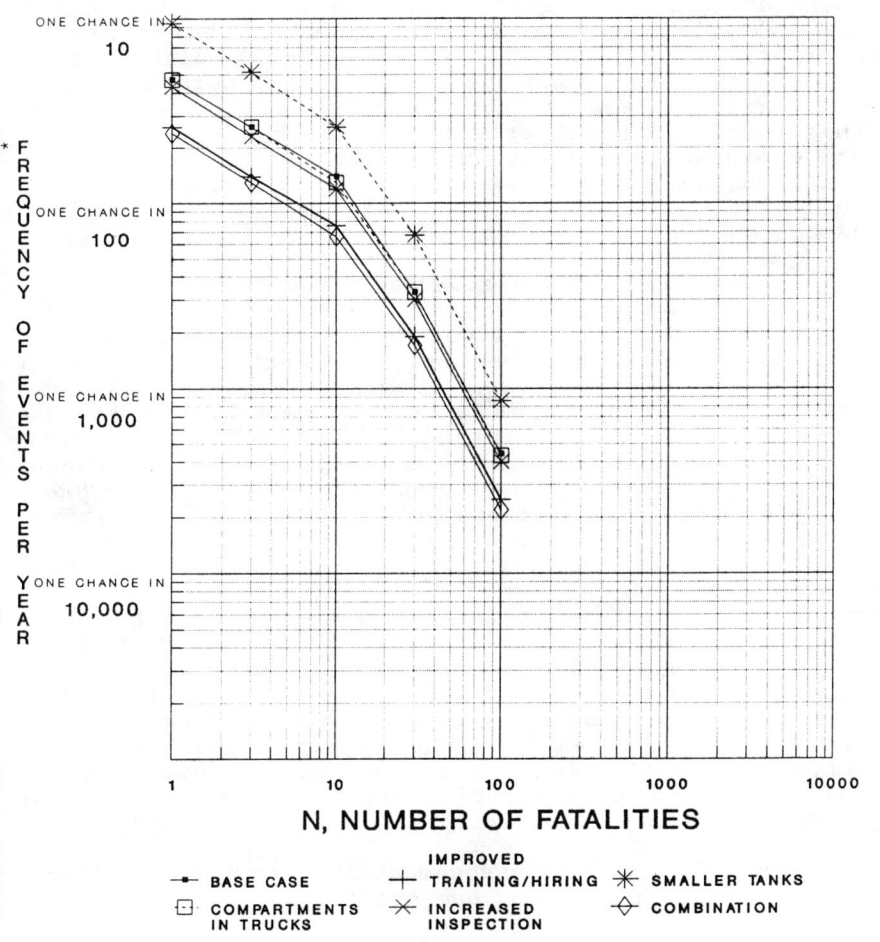

FIGURE 2
TRUCK TO LOS ANGELES
PROPANE - FATALITY
RISK REDUCING MEASURE ASSESSMENT

While many of these measures are very difficult to quantify, they can be evaluated in a qualitative manner. However, the effect that some measures might have on reducing overall risk can be quantified by reviewing accident rates for other roads that have undergone similar design modifications.

Once the set of risk reducing measures has been evaluated, it is possible to begin to estimate the combined effect of the various measures that could be incorporated into a HMT risk management plan. This estimate should include the measures that were evaluated quantitatively as well as those done qualitatively. The final result is a set of risk profile bands. The lower edge of these bands provide an estimate of the societal risk assuming a set of specific risk reducing measures. The bands also include an estimate of the uncertainty associated with the analysis.

(g) Ensure a publicly responsive process.

Undoubtly, public involvement in the process is essential. There may be several proven approaches to bring a meaningful public dialogue into the process. Whatever the approach, however, we emphasize that the earlier you begin the dialogue, the better your risk management program will be. Meaningful public dialogue equates with a two-way educational process. *If you choose to begin the public dialogue once a draft risk assessment is available for comment (as is often done), you've started too late.* Rather, the public dialogue should be initiated as soon as the local jurisdiction decides to prepare a risk assessment. In so doing, several benefits may result. For example, we quickly discovered that the public was often better equipped than we were to identify several route-specific hazards, particularly those that are temporal. Early dialogue also increased our sensitivity to emphasizing clear and succinct communication of risk. Lastly, early communication assists in better integrating societal values into the judgement of risk acceptability.

(h) Structure the risk assessment as a computerized model which can be updated and built upon.

The Santa Barbara County has developed a computerized transportation risk model that can be used to estimate the risk associated with various transportation modes, various transportation routes, and various hazardous materials. Figure 3 illustrates the way in which all the elements are pulled together to determine the risk associated with each mode and route. A number of these elements have been handled first outside of the computer model, and have then

FIGURE 3
RISK ANALYSIS COMPUTER MODEL APPROACH

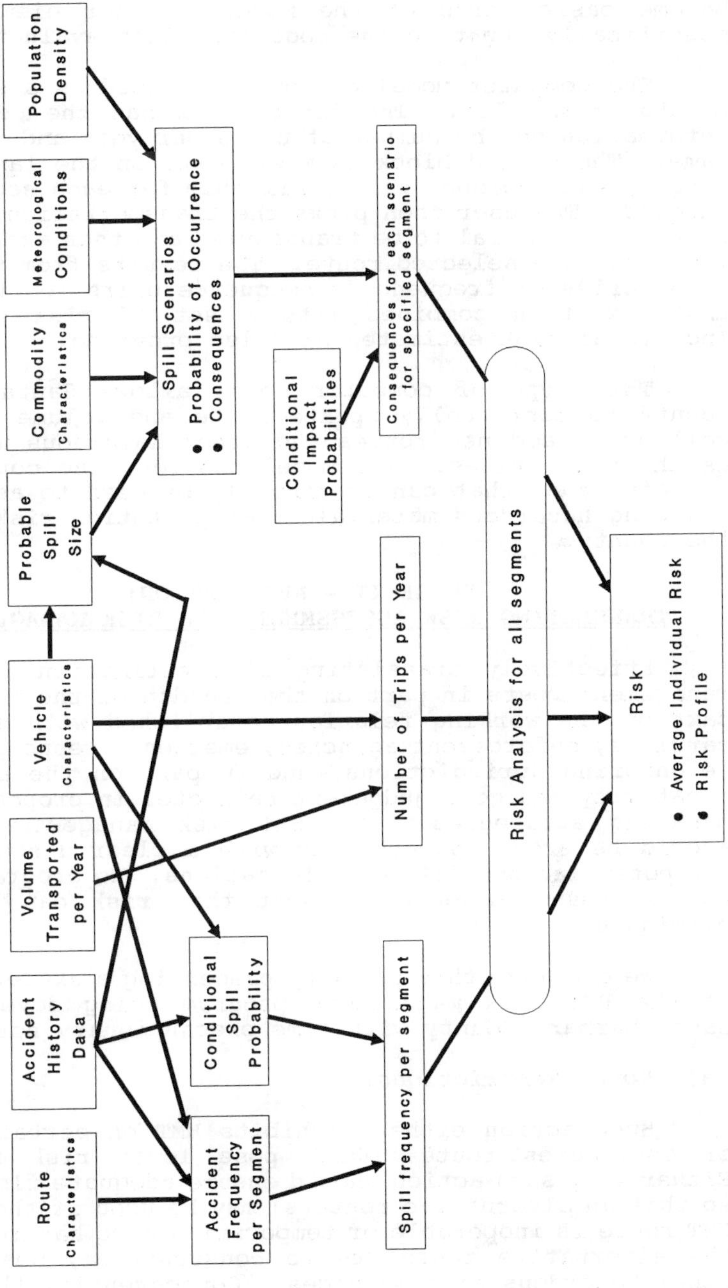

become basic parts of the model. Other elements are specifically input to the model for each evaluation.

The computer model can be visualized as a series of blocks or modules. The first block has the user input information on the number of trips per year and the route name. The second block is where data on the land usage, length, and accident rate applicable for each segment are entered. The user then picks the transportation mode and hazardous material to be transported and then executes the model for the selected route. The results from the model is a series of frequency/consequence pairs for the route that are then combined into a set of risk profiles. Individual risk estimates are also provided.

This type of computer model allows Santa Barbara County to continually update route and volume data, as well as to add new routes and other hazardous materials as the need arises. The model provides the county with a useful tool that can continually be used to assess the changing hazardous materials transportation risks within the County.

PHASE II - APPLICATION:
TRANSLATING RISK ASSESSMENT INTO RISK MANAGEMENT

Effectively translating risk estimation into risk management rests in part on the breadth of the former, in part on the working relation established with shippers, carriers, enforcement agencies, emergency responders, and neighboring jurisdictions, and in part on the amount of creativity and good judgement reflected in proposed risk-managing strategies. Any HMT risk management program should be presented publicly with a clear discussion of its potential benefits and limitations. No program should be so presented as to suggest that risk can be fully eliminated.

We conclude this paper by presenting a skeleton model of the HMT risk management program being developed in Santa Barbara County with some precautionary notes.

(a) Route Restrictions:

Such action either prohibits HMT on certain routes or designates routes which pose lower risk for HMT. Either way, such action should ensure adequate flexibility so that an alternative route(s) may be used if the primary HMT route is inoperable or temporarily more hazardous than the alternative route due to construction, weather, or other hazardous circumstances. Consequently, the action should consider temporary alternative routes and mechanisms to shift to alternative routes when so desired.

Additionally, attention should be given to notifying shippers and carriers of such restrictions. California requires that local jurisdictions post signs on the routes where HMT is prohibited.

(b) Route Improvements:

Routing restrictions, as mentioned previously, should be followed by measures to reduce hazardous conditions on designated routes. This avenue of risk management can include suver measures: better patrolling by enforcement agencies, better road maintenance, and correcting poorly designed road sections such as improperly banked curves, or by adding truck climbing lanes and runaway lanes on steep grades. Other measures to consider include mobile truck inspections, installation of call boxes for rapid reporting of accidents, and increased emergency response capabilities, particularly in more remote areas.

(c) Procedural and Technological Improvements:

Measures such as improved driver hiring and training, improved vehicle maintenance, and addition of truck safety devises, are the direct responsibility of the carrier. While some carriers should consider substantial improvements in these areas, others continue to set high standards of safe operation that far exceed minimum requirements. From a local perspective, we're interested in encouraging use of the latter. We have been working closely with shippers to carefully select those carriers with good safety programs. One shipper already had such a system developed, employing a comprehensive survey which addresses all facets of safe HMT risk management.

Additionally, we also work with large shippers to ensure visual inspection of both driver and truck tank before loading. The shipper assumes the responsibility of refusing to load a potentially unsafe vehicle.

As with any experimental program, our risk management program has many issues yet unresolved. We will continue to convene our HMT advisory committees and to facilitate public dialogue to systematically evaluate the effectiveness and efficiency of the program as it develops.

ASSESSING COMMUNITY SAFETY FOR HAZARDOUS MATERIALS TRANSPORT

Chung-Kuo Chiang, Ph.D., P.E.[1]
Edmund J. Cantilli, Ph.D., CSP, P.E.[2]
Stephen T. Ying, MSCE, P.E.[3]

ABSTRACT
This paper describes a computer model developed to assess the safety of a community through which hazardous materials will be transported. The model is predictive and can be used even in the absence of a past history of incidents. Instead, it draws on selective measures for both risk and community preparedness to meet that risk. Risk measures include various characteristics of drivers, vehicles, and roadways, and the exposure of population and property. Preparedness includes measures of emergency response capability, regulation compliance, public awareness, and evacuation plans. Developed for a LOTUS 1-2-3 spreadsheet, the model provides relative safety ratings and can guide communities in determining the most effective approaches for improving overall safety.

INTRODUCTION
As our industrial society's appetite grows for products and manufacturing processes that require the shipment of hazardous materials, so grow the amounts of hazardous materials passing through communities, the resulting accidents, and the need for some means of assessing community risk before accidents happen.

Defining a hazardous material as any material the United States Department of Transportation determines "to be in a quantity and form capable of posing an unreasonable risk to health, safety, or property when transported in

---
[1]Lead Engineer, Parsons Brinckerhoff Quade & Douglas, Inc., One Penn Plaza, New York, NY 10119
[2]Professor, Polytechnic University, Brooklyn, New York 11201
[3]Supervising Engineer, Parsons Brinckerhoff Quade & Douglas, Inc., One Penn Plaza, New York, NY 10119

commerce, and which has been so designated", gives nine classes of hazardous materials as listed in the Code of Federal Regulations, Title 49-Transportation, Parts 100 to 199 (CFR 49, 1987). Traveling by trucks, rail cars, water carriers, air carriers, and pipelines, such materials pass through virtually every community, the "machine" of the transport system interacting constantly with both man and the environment (Figure 1).

It has been estimated that at least 250,000 shipments of hazardous materials are made each day in the U.S., totaling at least 4 billion tons per year, and this volume is expected to double every ten years (Safety Effectiveness, 1981). The Department of Transportation estimates that between 5 and 15 percent of all trucks on the road at any time carry hazardous materials (Transportation of Hazardous Materials, 1983).

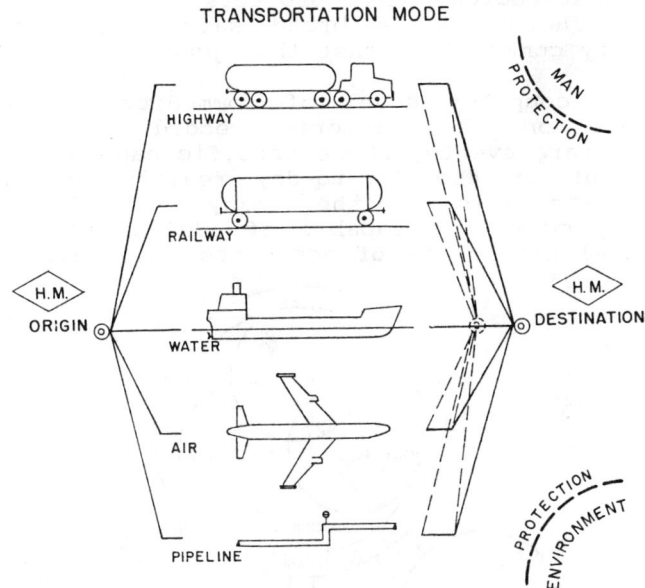

FIGURE 1   THE HAZARDOUS MATERIALS TRANSPORT SYSTEM

Risk Assessment as a Basis for Evaluating Overall Safety. In the face of such growing shipments, a computer model for assessing community safety should not be an "after-the-fact" model that depends for its input on data from accidents that have occurred. Instead it must predict at least relative levels of community safety based on factors observable before accidents occur. Such factors may be found in the interactions of the transport "machine" (vehicle) with both "man" and "environment". Here "man" is used in the sense of all those people who either operate the transport system (such as truck drivers) or are exposed to its dangers (such as roadside residents). "Environment" means on the one hand the total infrastructure of guideways, signs, markings, right-of-way, weather and light conditions, and police and emergency-response personnel, and on the other hand the communities, sensitive receptors, and natural surroundings to be protected from injury.

Risk factors are amenable to measure in advance of actual accidents. Their ratings provide a sort of feedback. A measurable change in some risk factor (or in community preparedness to meet that risk) affects overall safety. This sensitivity makes possible a risk assessment methodology that projects specific measures of various factors to a broad safety ranking whose acceptability communities must then judge (Figure 2).

The present computer model of community safety thus focuses not on the historic record of accidents (relatively rare events, whose specific causes are often obscure), but on the day-to-day reality of existing conditions. Its basis is the many hazards (often individually small but capable of adding up) that are simply causal components of accidents that have not yet happened.

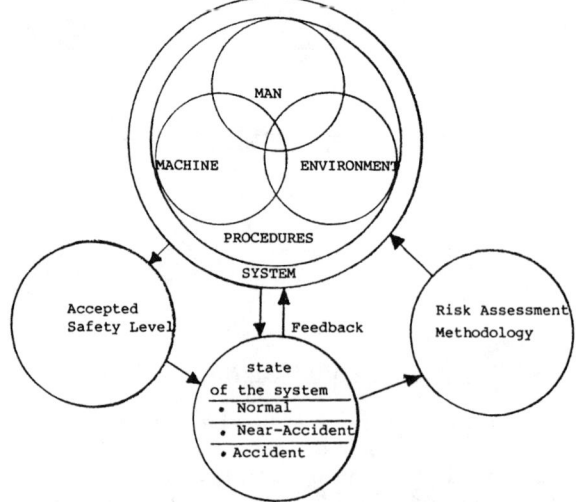

FIGURE 2: THE RELATIONSHIP OF RISK ASSESSMENT TO SYSTEM SAFETY (Cantilli and Horodniceani, 1980)

The Proposed Model. With community safety seen as a ratio of community preparedness to community risk, our model has been developed with two submodels. First is community risk (CR) such as the likelihood of an incident (fire, spillage, or explosion) and the relative exposure to death, injury, or property damage. Second is community preparedness (CP), emergency response capability and preparedness, such as personnel, equipment, public awareness, and evacuation plans. An equation for this is:

$$LOS = \frac{CP}{CR}$$

Where: LOS = Community Level of Safety
(Ranges from A to F)
CP = Community Preparedness Level Index
(Values from 1 to 100)
CR = Community Risk Level Index
(Values from 0.1 to 1.0)

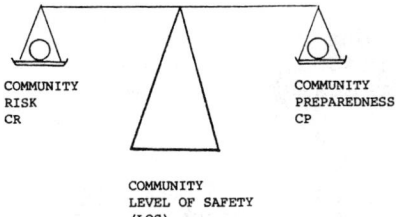

FIGURE 3: BASIC CONCEPT: EVALUATION OF THE COMMUNITY LEVEL OF SAFETY (LOS)

FIRST SUBMODEL: COMMUNITY RISK (CR)
This model focuses on roadway transport (where 90 percent of hazardous materials accidents occur), but its methodology can be adapted for other transport modes. In it, the first submodel is an evaluation of community risk based on three elements within some defined project area (often a half-mile to either side of a transportation corridor): the risk of a motor vehicle accident, the risk that such an accident will involve a carrier of hazardous materials, and the community exposure.

Risk Level of a Motor Vehicle Incident $R_L(MV)$. The risk of a hazardous material motor vehicle incident occurring to any given transport vehicle is measured not as a probability based upon past experience on a certain roadway, but as the potential for such an incident (fire, spillage, or explosion) arising from three contributing factors: the driver, the vehicle, and the roadway geometry/environment.

Assume then that the risk level of a motor vehicle accident/incident for all vehicles is proportional to traffic volume, the three above-mentioned individual contributing factors (driver, vehicle, roadway geometry/environment), and average vehicle speed. The combination of drivers, vehicles, and roadway geometry/environment level indexes contributes to the potential for accidents, and the average vehicle speed contributes to its severity. The principle of improving road safety becomes reducing the frequency and severity of accidents by taking suitable measures to reduce the total risk index.

Based on the above assumptions, therefore, the model for risk of a motor vehicle incident can be formulated as a relation among several risk factors, each rated, using professional judgment and various guidelines to promote consistent comparisons, on a scale of 1 to 10 (or normalized to that scale):

$$R_L(MV) = T_L \times (D_L + V_L + R_L) \times S_L$$

Where:
- $R_L(MV)$ = Risk level index of a motor vehicle incident. (Converted to values 1 to 10)
- T_L = Traffic volume level index.
- D_L = Driver level index.
- V_L = Vehicle level index.
- R_L = Roadway geometry/environment level index.
- S_L = Average vehicle speed level index.

Traffic volume (T_L) is rated 1 to 10 (Table 1).

TABLE 1   TRAFFIC VOLUME LEVELS (T_L)

Volume Level	Average Daily Traffic (ADT)	Roadway
1	5,000 and below	Local
2	5,000 - 10,000	Collector
3	10,000 - 15,000	Arterial
4	15,000 - 20,000	Rural Highway
5	20,000 - 25,000	Freeway
6	25,000 - 30,000	4 Lanes
7	30,000 - 40,000	Freeway
8	40,000 - 50,000	6 Lanes
9	50,000 - 60,000	Freeway
10	60,000 or more	8 Lanes

ASSESSING COMMUNITY SAFETY

The driver level index (D_L) is the number of the following 10 variables judged not satisfactory, based on driver interviews (typically 100 or more for a valid sample):

1. Proper licensing
2. Training
3. Experience as truck driver
4. Accident/violations history
5. Awareness of regulations
6. Awareness of emergency response actions
7. Knowledge of potential risk of hazardous materials carried
8. Knowledge of CHEMTREC hot line
9. Driving hours within limits
10. Character of driver and alcohol, drug involvement

Similarly, the vehicle level index, V_L, is the number of the following variables judged not satisfactory:

1. Proper registration
2. Physical condition
3. Braking system
4. Steering system
5. Condition of tires, suspension
6. Lights
7. Load security
8. Proper weight
9. Age of vehicle
10. Level of maintenance

Roadway geometrics/environmental level (R_L) can be determined as the average score of the following items evaluated on a scale from 1 (best) to 10 (worst):

1. Interchange on-off ramps or intersections
2. Horizontal curves
3. Vertical curves
4. Traffic control devices
5. Roadside hazards
6. Pavement condition
7. Type of median and shoulders
8. Unsafe road condition
9. Percentage grade
10. Unsafe weather condition

Average speed of vehicles (S_L) is rated:

10 = Average speed of vehicles over 60 MPH
 8 = 50-55 MPH
 6 = 40-45 MPH
 4 = 30-35 MPH

2   =   20 MPH or less

Risk Level of a Motor Vehicle Incident Involving Hazardous Materials R_L(HMV). The risk level of a hazardous materials vehicle incident simply assumes proportion to the overall motor vehicle incident:

$$R_L \text{ (HMV)} = R_L \text{ (MV)} \times (P_{EX} + P_{CG} + P_{FL} + P_{FS} + P_O + P_P + P_R + P_C + P_M)$$

Where:
- P_{EX} = % Explosives vehicles in ADT
- P_{CG} = % Compressed gas vehicles in ADT
- P_{FL} = % Flammable liquid vehicles in ADT
- P_{FS} = % Flammable solid vehicles in ADT
- P_O = % Oxidizers and organic peroxides vehicles in ADT
- P_P = % Poison vehicles in ADT
- P_R = % Radioactive vehicles in ADT
- P_C = % Corrosive vehicles in ADT
- P_M = % Miscellaneous or unknown class vehicles in ADT

Community Risk (CR). The community risk level (which will be converted to a scale from 0.1 to 1.0) is the above risk level of incidents involving hazardous materials times a factor representing the potential consequences. These consequences are estimated through five measures of potentially affected population, property, numbers of hazardous materials establishments (industry, storage facility, etc.), and numbers of sensitive facilities, multiplied by the number of forms of threat resulting from an incident, that is:

$$CR = R_L \text{ (HMV)} \times (P_D + P_V + N_E + N_S) \times T_F$$

Where:
- R_L (HMV) = Risk level index of an incident involving a motor vehicle carrying hazardous materials
- P_D = Population density factor
- P_V = Property value factor
- N_E = Condition and number of hazardous materials establishments
- N_S = Type and numbers of sensitive facilities (schools, hospitals, etc., in the vicinity of transport routes)
- T_F = Forms of threat resulting from an incident: fire, spillage, explosion, or combination

The above factors contributing to the consequences of a hazardous materials accident/incident are rated as follows:

- **Population Density Factor**

Average Population Density per Mile^2	Rating
<100 population	2
100 - 500	4
500 - 1,000	6
1,000 - 10,000	8
>10,000	10

- **Property Value Factor**

Average Property Value per Mile^2	Rating
<$1 million	2
$1 - 5 million	4
$5 - 10 million	6
$10 - 100 million	8
>$100 million	10

- **Number of Hazardous Materials Establishments Factor**

Average Numbers per Mile^2	Rating
<2 establishments	2
4 - 6	4
6 - 8	6
8 - 10	8
>10	10

- **Number of Sensitive Facilities Factor**

Average Numbers per Mile^2	Rating
<2 sensitive facilities	2
4 - 6	4
6 - 8	6
8 - 10	8
>10	10

- **Forms of Threat Factor.** From road surveys and visual identification of placards on vehicles, the peak one-hour count is observed for vehicles carrying the various types (placard groups) of hazardous materials. These are then weighted according to Table 2 to allow for the fact that some placard groups pose more forms of threat than others (exploding as well as burning, for example), and translated to ratings (2-10).

TABLE 2
FORMS OF THREAT RESULTING FROM AN INCIDENT

Type of Hazardous Placard Group	Form of Threat			Weight of Placard Group
	Fire	Toxic Spill	Explosion	
	1	2	3	
Explosives			X	3
Compressed Gas	X		X	4
Flammable Liquid	X	X	X	6
Flammable Solid	X			1
Oxidizer and Organic Peroxides		X	X	5
Poison	X	X		3
Corrosive		X		2
Radioactive		X		2
Miscellaneous	X			1

Total Adjusted Placard Count	Rating
<50	2
50 - 100	4
100 - 150	6
150 - 200	8
>200	10

SECOND SUBMODEL: COMMUNITY PREPAREDNESS (CP)

The second submodel in this model is an evaluation of community preparedness (CP), seen as the sum of the Emergency Response capability index (ER) and the Regulation Compliance index (RC).

$$CP = ER + RC$$

Where:
- CP = Community Preparedness level index.
  (A value range from 1 to 100) - addition of ER and RC values.
- ER = Emergency Response Capability index.
  (A value range from 0 to 50 - converted from average rating score.)
- RC = Regulation Compliance index.
  (A value range from 0 to 50 - converted from average rating score.)

Emergency Response Capability Index (ER). ER is the average value of ratings of the following 10 variables, each rated on a scale of 1 (worst) to 10 (best) based on the evaluator's best knowledge and professional judgment.

1. Manpower available for hazardous materials transportation accident/incident emergency response
2. Training level of available personnel
3. Communication capability
4. Access and transportation to the emergency scene
5. Emergency response time from the incident occurrence to clean-up
6. Quality and amount of needed equipment
7. Public awareness (through media) and capability to cope with emergency
8. Community preplanning programs for the anticipated or potential risks
9. Capability to implement those programs effectively, including practical exercises
10. Capability to evacuate people seriously exposed to the emergency

Regulation Compliance (RC) Index. Similarly, the Regulation Compliance index (RC) is an average score on a scale of one (worst) to ten (best) for the ten variables below, based on the evaluator's best knowledge and professional judgment:

1. Community History of Release/Incident
2. Government Penalties for Shipper/Transporter Violation
3. Level of Government Inspection
4. Hazardous Materials Transport Routes
5. Extent of Shipper/Transporter Violations
6. Condition of Shipping Papers
7. Warning Placards on Vehicles
8. Packaging of Hazardous Materials Cargoes
9. Hazardous Materials Quantity Limitations
10. Manufacturing, Storage Safety

OVERALL RATING: COMMUNITY LEVEL OF SAFETY (LOS)
The Level of Safety of a community can be rated from letter A to F according to the ratio of community preparedness to risk (Table 3). Level of Safety C or better is considered acceptable by most communities, with D the minimum acceptable level in some cases, and E and F unacceptable. The lower-ranked communities need to improve their level of safety by increasing preparedness, reducing risk, or both.

Also in Table 3, individual values of community preparedness (CP) and community risk (CR) are shown in their approximate correlation to the level of safety, even though only their ratio is in fact used by the model for assigning the A through F ratings. Nonetheless, unusually high risk or low preparedness is in itself unacceptable, whatever the ratio may work out to be. This fact is made readily visible in Figures 4 and 5, where specific values are shown as unacceptable in our professional judgment, acceptable, or at an in-between level where risk reduction or preparedness improvement is desirable.

TABLE 3 COMMUNITY LOS (LEVEL OF SAFETY)

LOS	CP/CR	CP	CR	DESCRIPTION
A	760-1000	95-100	0.100-0.124	Community with extremely high preparedness and extremely low risk.
B	500- 759	80- 94	0.125-0.159	Community with high preparedness and low risk.
C	300- 499	60- 79	0.160-0.199	Community with medium preparedness and medium risk.
D	125- 299	40- 59	0.200-0.319	Community with low to medium preparedness and medium to high risk.
E	40- 124	20- 39	0.320-0.499	Community with very low preparedness and very high risk.
F	1- 39	1- 19	0.500-1.000	Community with extremely low preparedness and extremely high risk.

ASSESSING COMMUNITY SAFETY   145

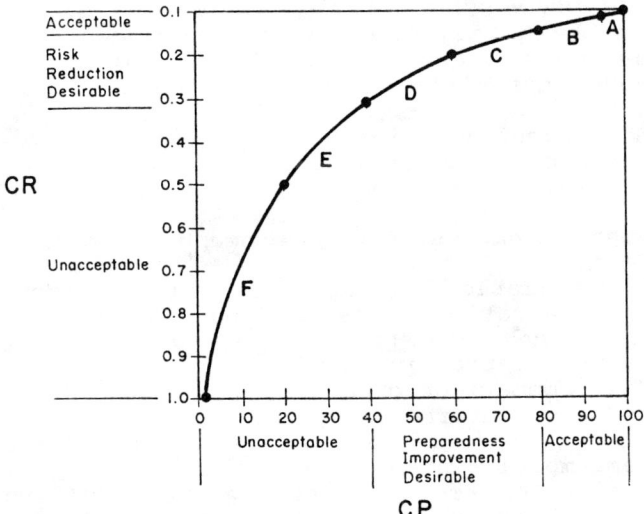

FIGURE 4: LOS VS. CR VS. CP

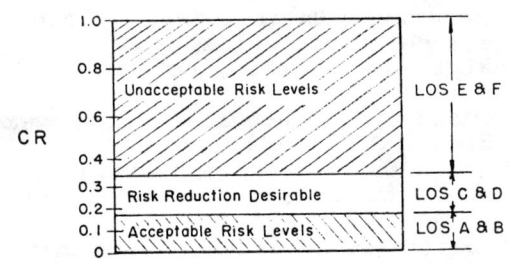

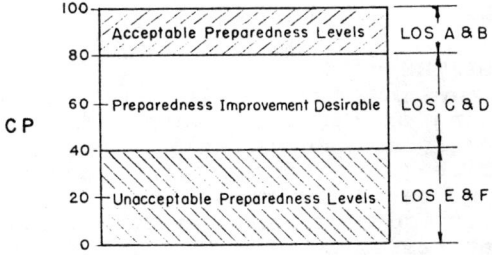

FIGURE 5: RANGE FOR SETTING ACCEPTABLE LEVELS

USERS OF THE MODEL
It has been suggested that there would be three categories of users interested in applying a model of hazardous materials transport (Russell and Smaltz, 1986):
o   Government Agencies
o   Hazardous Materials Establishments
o   General Public (Communities)

Government Agencies. For government agencies, the model provides:
o   A systematic uniform basis for the identification and evaluation of risks, both the potential chances and the consequences of hazardous materials transportation incidents
o   Risk comparison of a commodity shipped in different modal or intermodal systems
o   Risk comparison for different commodities in the same modal system
o   Evaluation of different regulatory alternatives to show which would result in economic inefficiencies, dislocations, and waste

Hazardous Materials Establishments. These organizations can use the model formulation:
o   To analyze and identify potential hazards and their consequences
o   To promote and reduce risks in advance of design or operational decisions
o   To optimize operations and minimize costs
o   To improve the safety of existing systems

General Public (Communities). Communities can assess their own:
o   Community preparedness or vulnerability
o   Emergency response capabilities
o   Community levels of safety

BENEFITS OF THE MODEL
The purpose of this model is to get a clearer understanding of hazardous materials transportation and achieve the following goals:
o   Describe a community's hazardous materials transportation problem
o   Develop support and justification for community budget requests
o   Evaluate community enforcement program
o   Evaluate or provide adequate training/prevention programs
o   Plan future response/protection needs
o   Provide proper allocation of resources

REFERENCES

Cantilli, E. J., and M. Horodniceani, "Transportation System Safety Guidelines," Polytechnic Institute of New York, Brooklyn, New York, 1980.

<u>Code of Federal Regulations (CFR), Title 49-Transportation</u>, United States Government Printing Office, Washington, D.C., 1987.

Russell, E.R., and J. J. Smaltz, "A Community Model for Handling Hazardous Materials Transportation Emergencies," Vols. I, II, III & IV, USDOT, January 1986.

<u>Safety Effectiveness Evaluation: Federal and State Enforcement Efforts in Hazardous Materials Transportation by Truck</u>, National Transportation Safety Board, Washington, D.C., February 1981.

<u>Transportation of Hazardous Materials: Toward a National Strategy (Vols. 1 & 2)</u>, Transportation Research Board (TRB) Special Report 197, Washington, D.C., 1983.

Highway Robbery:
The Social Costs of Hazardous Materials Incidents on the Capital Beltway

Theodore S. Glickman, Molly K. Macauley and Paul R. Portney[1]

Abstract

In the summer of 1988, three major truck incidents involving hazardous materials occurred on the Capital Beltway in Washington, D.C. This paper estimates the delays that were incurred and the social costs of the incidents, which ran into millions of dollars. The impacts of policy options for reducing such delays and the associated costs are analyzed in the context of the three incidents.

Introduction

In the late summer of 1988, three major incidents involving hazardous materials occurred on the Capital Beltway in the space of less than a month. Together, they resulted in one fatality, thirteen injuries (two truck drivers, three motorists, and eight firefighters), and hours of delay to hundreds of thousands of vehicle occupants. For months afterward, media attention focused on the problem as public representatives and safety experts debated the solution and finally acted by restricting hazmat trucks to the two rightmost lanes. This paper takes a closer look at those incidents by estimating the delays that were created and the social costs that were incurred, and then observing that the cost of delay could have been reduced substantially if it had been factored into emergency plans and incident management decisions.

The Capital Beltway

Opened in August of 1964, the Capital Beltway is a 63-mile long link in the Interstate highway system. It was originally intended to serve as a bypass around Washington, D.C. for long-distance travellers and, coincidentally, to provide an efficient evacuation route in times of national emergency. Its first traffic jam took place on opening day, as eager users were backed up for miles while the ribbon-cutting ceremony was held near the

---

[1] Center for Risk Management, Resources for the Future, 1616 P Street, N.W., Washington, DC 20036.

New Hampshire Avenue exit. Since then, as the Washington suburbs have grown, it has become a major commuting artery, handling about 600,000 vehicles per day. Six to ten percent of that volume is trucks, which are involved in 17% of the accidents. On an average day, one of the six traffic accidents on the Beltway involves a tractor trailer. In a recent 16-month period, 4 of the 13 major truck accidents on the Beltway involved hazardous materials tankers.

The decision to impose the two-lane hazmat restriction following the three incidents in question was not universally supported. Its detractors were concerned that it would tend to increase the density of hazmat tank trucks in the occupied lanes and introduce conflicts with exiting traffic, while its supporters argued that it would help to alleviate the safety concerns associated with having small, maneuverable passenger cars occupy the same lanes as large, less maneuverable hazmat tank trucks. In another move made in the wake of these incidents, the Pentagon discontinued its fuel shipments during rush hours on the Beltway and 38 other federal agencies at 120 locations in the metropolitan Washington area followed suit. Other proposals have been made to ban all hazardous materials trucking on the Beltway during rush hours (which would divert tank trucks to other roads) and to introduce centralized tracking of all hazardous shipments (which would be expensive). The American Automobile Association [1989] has recommended that any trucker causing an accident on the Beltway be fined an amount based on the level of delay that is created and that the proceeds be used to fund public information programs to improve highway safety.

In the future, the Beltway will be equipped with a driver information system whereby incidents will be detected with television cameras and instructions will be relayed to motorists electronically. Plans have also been proposed for carpool lanes that would become part of a regional high-occupancy vehicle (HOV) network and for Interstate bypasses to the east and west of the Beltway. Considering the projections of 300,000 vehicles a day on some of its 8-lane segments by the year 2010, Beltway drivers will need all the help they can get.

Summary of The Incidents

The first of the three 1988 incidents of interest took place at 3:20 pm on Friday, August 12 on the outer loop of the Beltway near Route 193 (Georgetown Pike) in McLean, Virginia. A truck carrying about 10 tons of potentially explosive potassium permanganate in powdered form caught fire while bound for a Fairfax water treatment plant, causing the entire Beltway to be closed down until 6:45 pm, when the inner loop was reopened. The outer loop was kept closed until 9:15 pm. Eight firefighters were injured and more than 70 persons were evacuated. Traffic in downtown Washington was reportedly slowed by the incident.

On Thursday, August 26, the second incident occurred at 3:55 pm when a gasoline tanker on the inner loop hit the rear of a van that was attempting to pass the car in front of it, then crossed the concrete median wall and burst into flames. The location was New Carrollton, Maryland, just north of the Baltimore-Washington Parkway. A motorist travelling on the outer loop was killed when his car ran into the burning truck. Both sides of the Beltway were closed as firefighters spent more than two hours putting out the flames from the 3,000 gallons spilled. The inner loop was reopened at 8 pm and two of the four lanes in the outer loop were re-opened at 11 pm. The remaining lanes, 400 feet of which were melted by the fire, were repaved overnight. The drivers of the tanker and the van were both hospitalized, as were two other passengers in the van.

The third incident took place on Wednesday, September 7 at 11:05 am when a gasoline tanker overturned on an entry ramp on the inner loop in Annandale, Virginia, at the intersection of Route 236 (Little River Turnpike), injuring the driver. The ensuing flames, which damaged the steel beams of the overpass, were put out by 12:30 pm, but the outer loop was not reopened until 2:10 pm and the inner loop was kept closed until 5:30 pm, when 3 of the 4 lanes were reopened. Pavement damage in the remaining lane was repaired overnight. The ramp has a high level of tank truck activity because of its proximity to a tank farm in Fairfax City and was previously the scene of a similar accident that proved fatal to the driver, who was apparently taking the turn too fast.

Delay Estimation

Associated with each incident is a total delay in each direction on the Beltway, as measured by the number of vehicle-hours of waiting time. The Federal Highway Administration has developed an analytical procedure for computing the delay during any kind of freeway incident that reduces the normal capacity of a road for some length of time (see Morales [1986]). We used the FHWA model to estimate the delay for each of the incidents, given various sources of data on the traffic volumes, the roadway capacities, and the lane closure times.

The FHWA model assumes that there is a given freeway capacity S_1 which is reduced to an initial bottleneck capacity of S_3 during the time T_1 needed to detect the incident, followed by a capacity of zero during the time T_2 needed to respond to the incident and then an adjusted bottleneck capacity S_4 during the time T_3 needed to clear the lanes. From then on, the capacity is assumed to return to S_1. Any one or two of the times T_1, T_2 and T_3 can be zero. On the demand side, the initial level is S_2, which lasts for a time T_4 (which may be zero), after which the level changes to S_5. These parameters fully determine the total delay.

To estimate the demand flow levels, we obtained hourly traffic counts from the Virginia Department of Transportation for the appropriate days of the week in the preceding July near the locations of the first and third incidents. Similar data could not be obtained for the Maryland location, so we assumed that Virginia data for the first location also applied to the second incident. To simplify the calculations, we determined the average hourly traffic volume for the "heavy" period from 6 am to 10 pm and the "light" period from 10 pm to 6 am for each of the two datasets, using the counts for the appropriate days of the week and making separate determinations for the inner and outer loops of the Beltway. Figure 1 shows the traffic counts that we used and the dotted lines show the averages for the heavy and light periods.

Table 1. FHWA Model Parameters for the Three Incidents

	August 12		August 25		September 7	
	Inner Loop	Outer Loop	Inner Loop	Outer Loop	Inner Loop	Outer Loop
S_1	7400	7400	7400	7400	7400	7400
S_2	5219	4820	5278	5262	5657	5714
S_3	0	0	0	0	0	0
S_4	0	0	0	2700	1300	0
S_5	5219	4820	5278	1013	1310	5714
T_1	0	0	0	0	0	0
T_2	205	355	245	425	385	185
T_3	0	0	0	360	690	0
T_4	0	0	0	365	655	0

Note: S values are in vehicles per hour and T values are in minutes.

Table 1 contains the values of the parameters for the six model runs that we made to estimate the delays and Table 2 summarizes the results. In every case, in the absence of information to the contrary, we assumed that S_3 and T_1 are zero, i.e., that every lane was blocked immediately. The values of S_2 and S_5 in Table 1 are the appropriate averages from the traffic count data and the values of S_4 are taken from Urbanek and Rogers

Figure 1A

Hourly Traffic Counts on the Beltway
Route 193 to G.W. Parkway

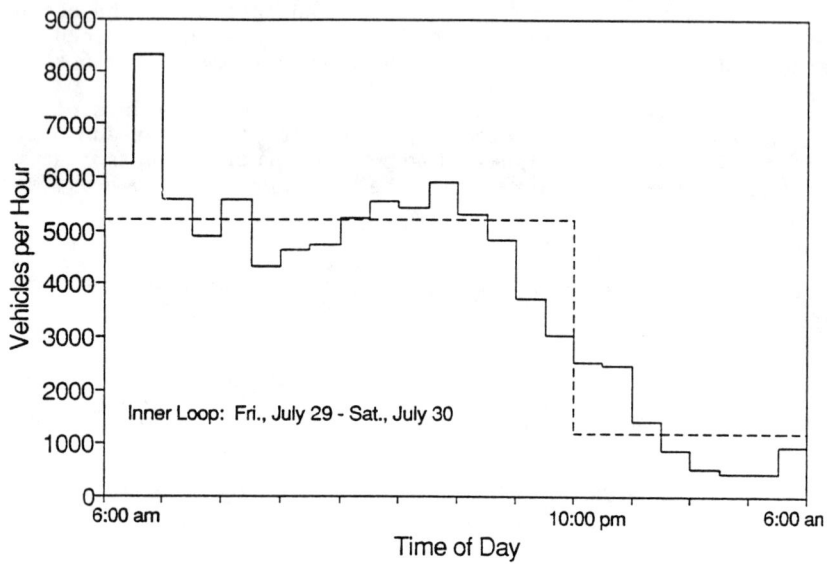

Inner Loop: Fri., July 29 - Sat., July 30

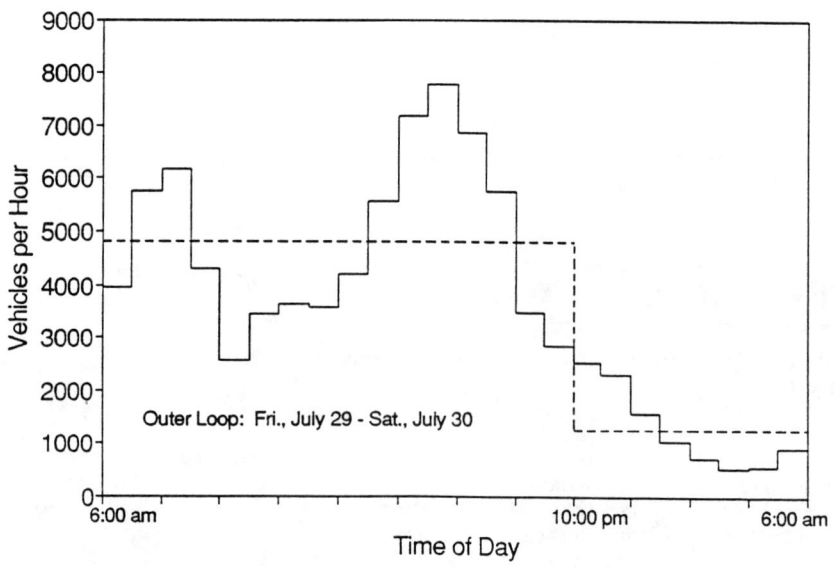

Outer Loop: Fri., July 29 - Sat., July 30

Figure 1B

Hourly Traffic Counts on the Beltway
Route 193 to G.W. Parkway

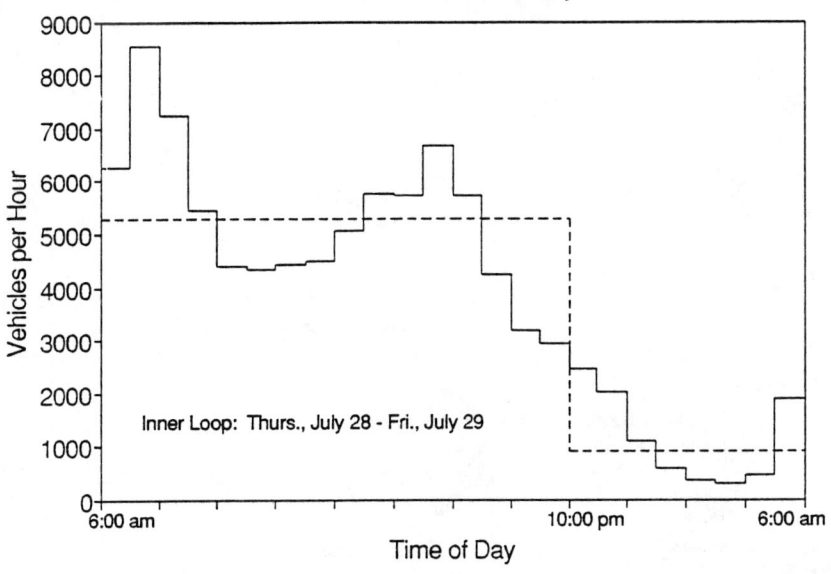

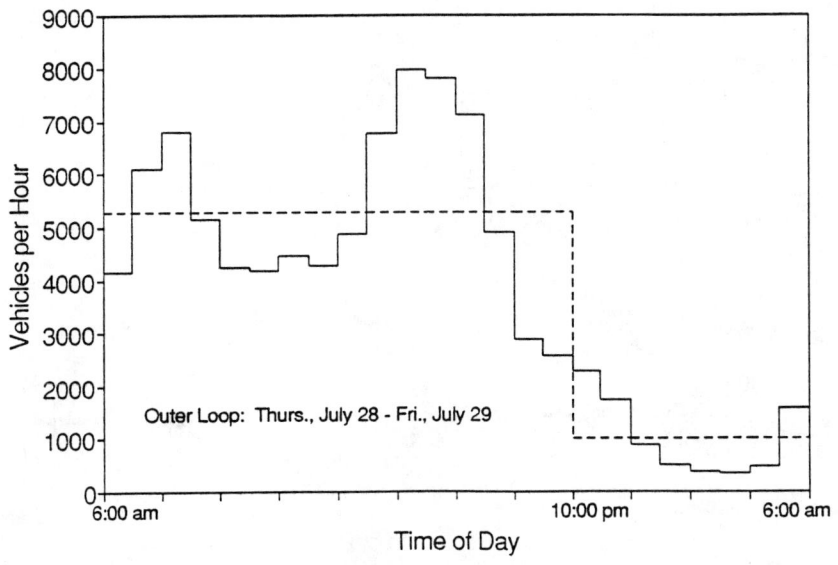

Figure 1C

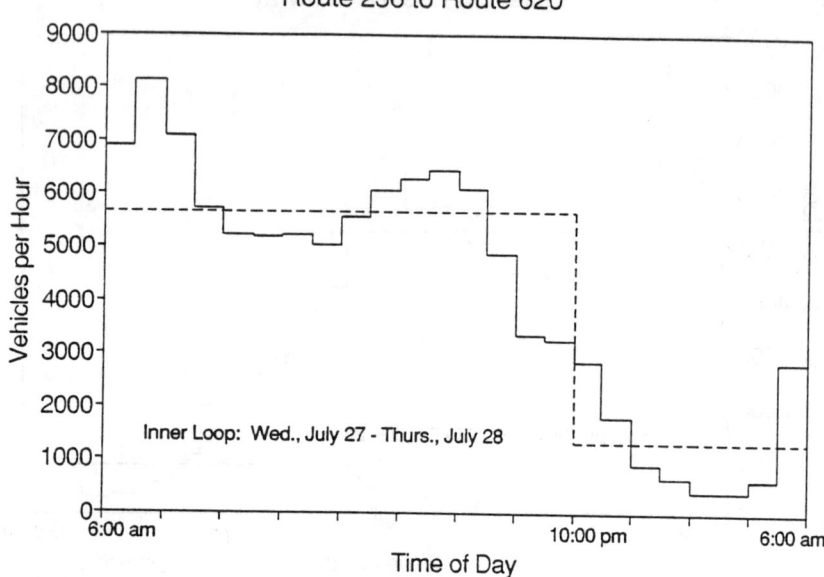

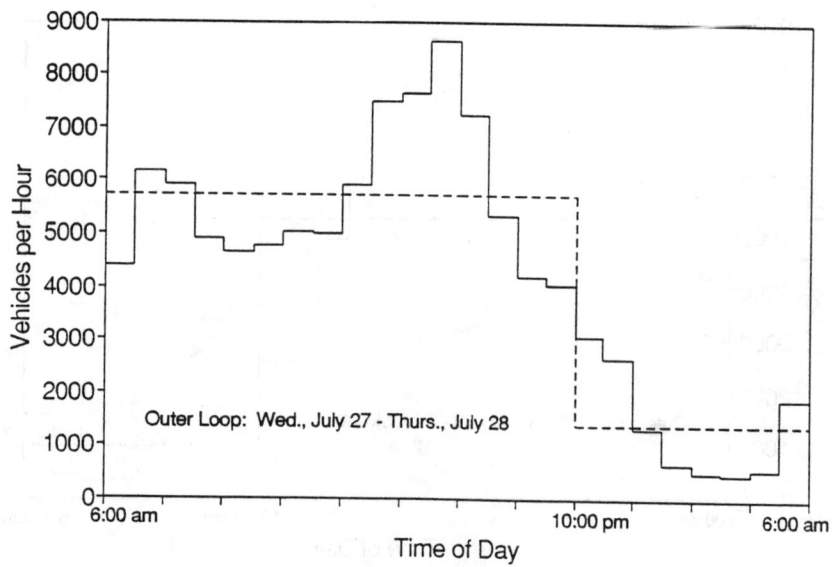

[1977]. In addition to the total delay, the FHWA model estimates the time to normal flow (TNF), which is the time between the onset of the incident and the moment at which the delay stops accumulating. Table 2 reports these estimates, along with the estimated delay per vehicle, which we found by dividing the total delay by the total demand during the TNF in each case.

Table 2. FHWA Model Results for the Three Incidents

	August 12	August 25	September 7
Total Delay (veh-hrs)			
Inner Loop	103,360	153,450	972,750
Outer Loop	241,980	339,720	119,210
Combined	345,340	493,170	1,091,960
Time to Normal Flow (hrs)			
Inner Loop	11.6	14.2	27.1
Outer Loop	17.0	16.7	13.5
Total Demand (vehs)			
Inner Loop	60,510	75,160	82,970
Outer Loop	81,800	42,730	77,330
Total	142,310	117,890	160,300
Delay per Vehicle (hrs)			
Inner Loop	1.7	2.0	11.7
Outer Loop	3.0	7.9	1.5
Combined	2.4	4.2	6.8

The results in Table 2 show that the total delay ranged from about 350,000 vehicle-hours in the first incident to about 500,000 vehicle-hours in the second and over 1,000,000 vehicle-hours in the third. On a per vehicle basis, the average delay ranged from 2.4 hours in the first incident to 4.2 hours in the second and 6.8 hours in the third. Note that we have assumed in these calculations that vehicles were not free to leave the Beltway once the incident began, hence these numbers are likely to be overestimates. However, given that we have not accounted for the delays that each incident induced on other roads, the combined vehicle-hours of delay in this table may in fact be underestimates of the systemwide delay for each incident.

Figure 2 shows how the capacity, demand flow and bottleneck flow accumulate over time in each direction during each incident, according to the runs of the FHWA model. The area enclosed by the demand flow and bottleneck flow lines is equal to the total delay in each case.

Social Costs of the Incidents

In estimating the social costs of the three incidents, we accounted for: (1) the actual direct costs, and (2) the imputed costs of delay. Information on some of the direct costs was gathered from a variety of sources, including local fire departments, local and state transportation departments, and trucking firms. Table 3 summarizes this information under three categories: emergency response, clean-up, and truck and lading loss. Emergency response costs refer to the value of the time of hazmat units, fire personnel, and traffic controllers, and the equipment and supplies they used (strictly speaking, some of these costs would have been incurred even if no incidents had happened). Clean-up costs are expenses for wreckage clearance, spill control and removal, and road repair and repavement. In the case of the third incident, we were only able to obtain a combined total for emergency response and clean-up (the reason this figure is so relatively high is that the damaged bridge had to be repaired). The costs of truck and lading loss, the third category, refer to the value of the truck cabs and trailers that had to be replaced and the cargoes that were lost.

Table 3. Estimated Social Costs of the Three Incidents

	August 12	August 25	September 7
(1) Direct Costs ($)			
Emergency Response	10,717	4,900	*
Clean-up	40,000	60,081	*
Subtotal	50,717	64,981	237,123
Truck and Lading Loss	124,000	103,000	108,000
Extra Fuel Consumption	200,877	286,039	633,337
Total Direct Costs	375,594	454,020	978,460
(2) Cost of Delay ($)	4,195,840	5,974,656	13,228,877
Total (1) + (2)	4,571,434	6,428,676	14,207,337
Ratio (2) ÷ (1)	11.2	13.2	14.5

*Breakout not available.

INCIDENTS ON THE CAPITAL BELTWAY

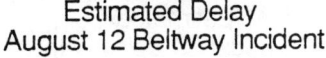

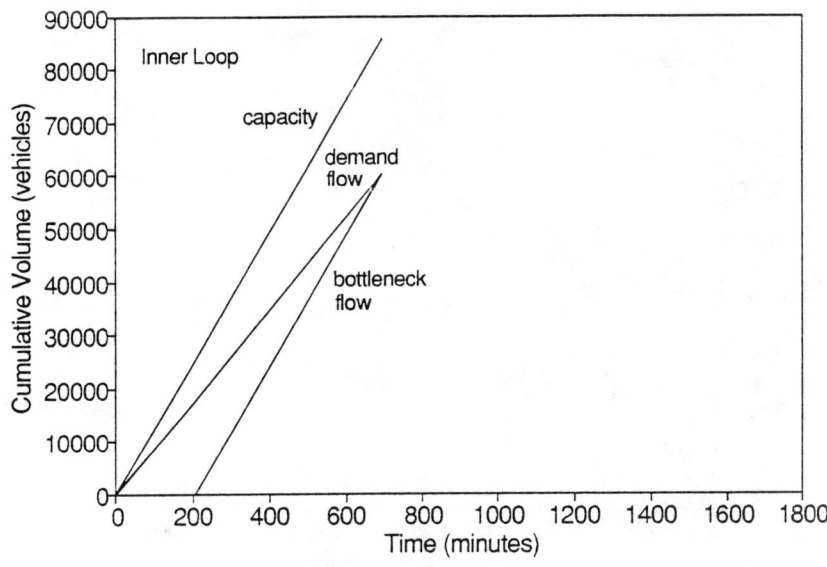

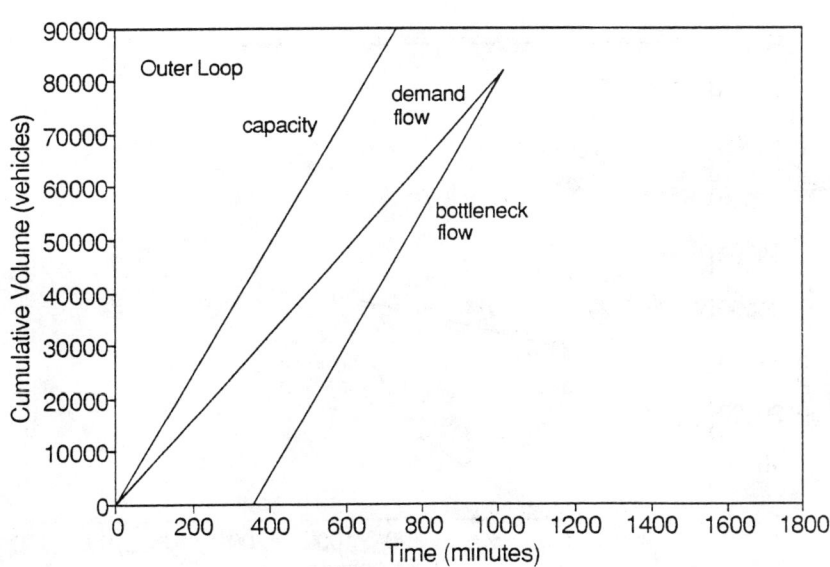

Figure 2B

Estimated Delay
August 25 Beltway Incident

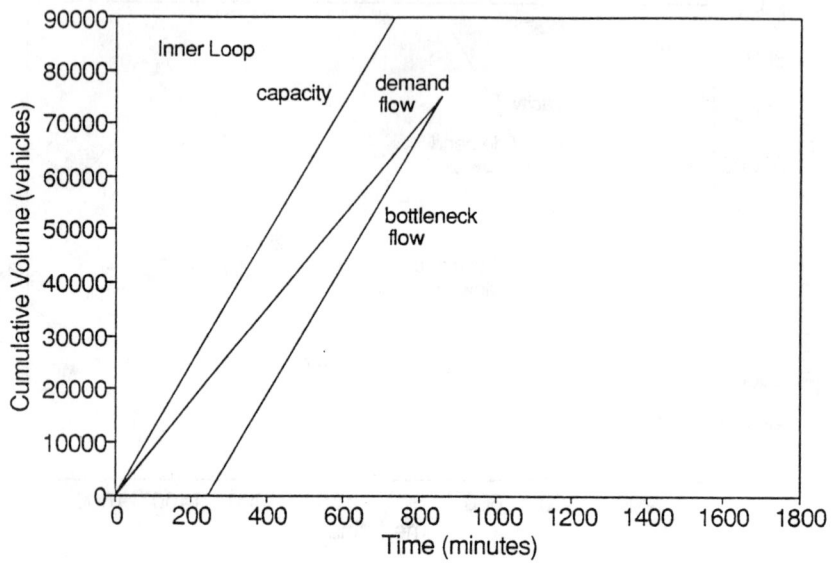

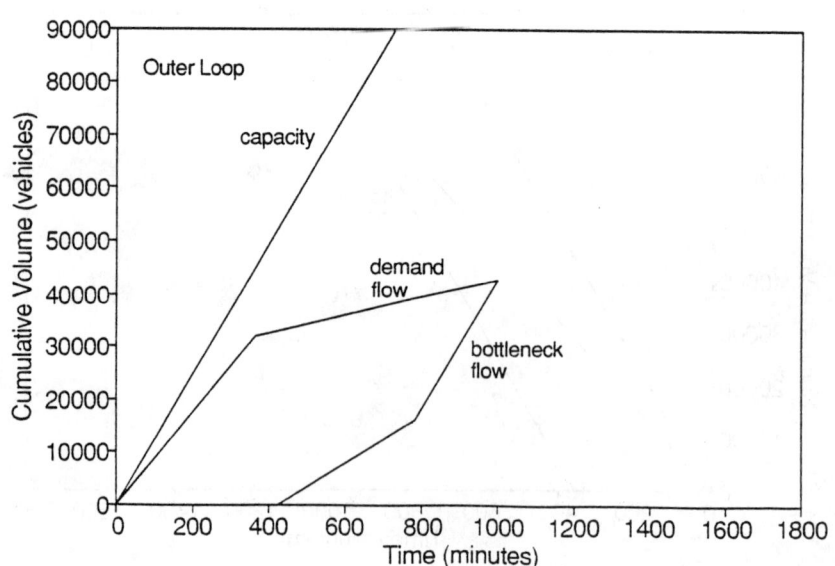

INCIDENTS ON THE CAPITAL BELTWAY

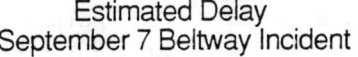

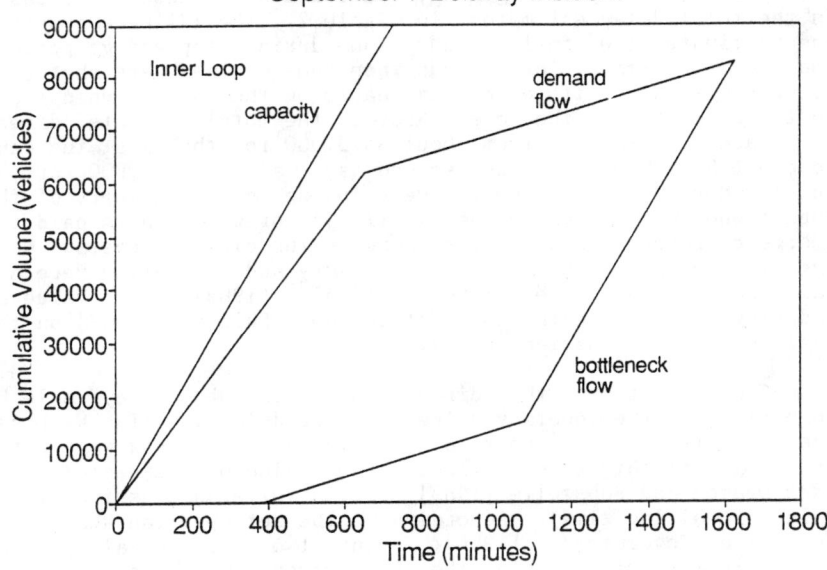

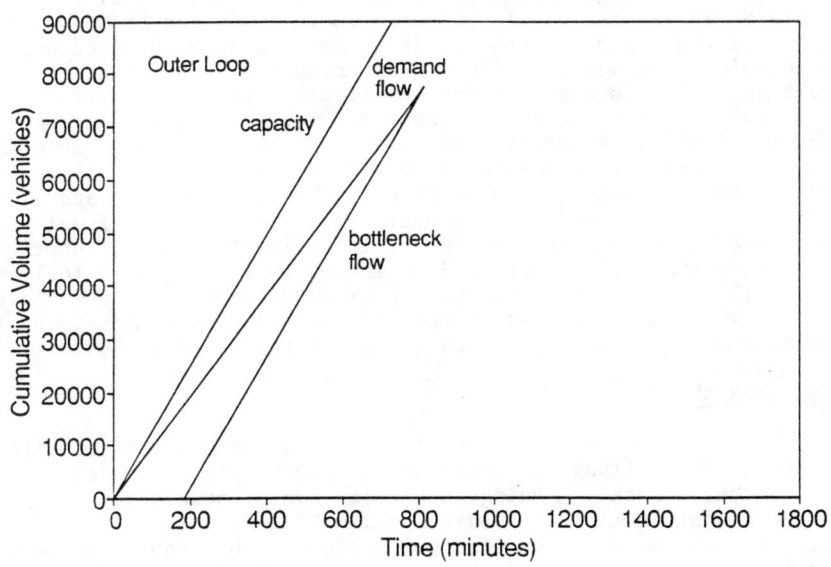

Also under the heading of direct costs are the extra fuel consumption costs for the delayed vehicles. Assuming that all the engines were kept running the entire time in each incident, that idling engines consumed 0.58 gallons per hour, and that the gasoline cost $1.00 per gallon, we calculated these costs based on the total delay estimates in Table 2. The idling assumption underestimates the fuel consumption during stop-and-go periods and overestimates it for periods when some engines were shut off, so we would expect it to be reasonable on the whole. When these costs are added to the ones above, the total estimated direct costs are seen to vary from about $375,000 for the first incident to about $455,000 for the second and almost $980,000 for the third. Most of these costs were borne by the members of the public who were caught in the traffic jam or whose taxes paid for emergency response and at least some of the clean-up costs. Note that the social costs of death, injury and evacuation were not taken into account (see Fisher, et al. [1989] on the value of reducing risks of death and Witzig and Shilleen [1987] on the evaluation of evacuation costs).

Significant as the direct costs may be, they pale by comparison to the monetary value of the delay itself. We based the estimates of this cost on an average after-tax wage rate, assuming that this rate reflects the value of unexpected delay (see Deacon and Sonstelie [1985] for a fuller discussion of the value of waiting time). According to the Metropolitan Washington Council of Governments [1986], about two-thirds of all Beltway travel is for trips within the D.C. area. Thus, we assumed an average hourly wage rate of 2/3($9.77) + 1/3($8.06) = $9.20, where $9.77 was the local wage rate in 1988 and $8.06 was the national wage rate. Further, we assumed an automobile occupancy rate of 1.24 persons (the 1985 rush-hour estimate for the Beltway, which is close to the 1.3 figure employed by Teal [1988] as a national average for all roads). The results of multiplying the total delay estimates from Table 2 by this wage rate and occupancy rate are as shown in Table 3: about $4.2 million for the first incident, $6.0 million for the second, and $13.2 million for the third. Hence, the total cost of delay is estimated to have exceeded the total direct costs of the three incidents by as little as 11.2 to 1 and as much as 14.5 to 1, or about 13 to 1 on the average. Of course, this ratio could be even higher if successive increments of delay were considered to be increasingly burdensome, as suggested by Larson [1988].

Reducing the Cost of Delay

The finding that the cost of delay dominates the direct costs of the Beltway incidents by a wide margin raises the issue of whether enough is being done to avoid such delays. Three general approaches to avoiding delay immediately come to mind: (a) shorten the duration of the incident; (b) reduce the demand during the incident; and (c) increase the capacity during the incident. The first approach requires rapid detection, response and/or clearance. The second approach requires improved diversion of traffic away from the incident scene. The third approach, which is more controversial, requires that fewer lanes be closed or that some lanes be opened sooner (or, at least, that a shoulder be opened up to let some traffic through).

By changing the values of the parameters in Table 1, we conducted a sensitivity analysis that uses the FHWA model to show what impacts each of these approaches would have had on the delays (and the costs of delay) estimated for the three incidents. First we reduced all T_2 and T_3 values by 10 percent to shorten the incident duration. Then we reduced all S_2 and S_5 values instead by 10 percent to lessen the demand rate. Finally, to represent the situation where the far right lane would be kept open throughout the incident on the opposite side of the road, we increased the respective value of S_3 from 0 to 1300 (the capacity when one lane is open), reduced T_2 to zero and replaced T_1 with the former value of T_2.

Table 4. Results of the Sensitivity Analysis

	August 12	August 25	September 7
Reduction in Delay (%)			
(a) Shorten Duration by 10%	19.0	14.2	15.1
(b) Reduce Demand by 10%	25.1	18.4	19.1
(c) Increase Capacity by One Lane (opposite side only)	11.4	11.8	4.0
Saving in Cost of Delay ($M)			
(a) Shorten Duration by 10%	0.75	0.80	1.88
(b) Reduce Demand by 10%	0.99	1.04	2.38
(c) Increase Capacity by One Lane (opposite side only)	0.45	0.67	0.50

Table 4 is based on the results of rerunning the model with these changes. It shows that fairly modest improvements in emergency management would have yielded substantial benefits, reducing the number of vehicle-hours of delay in an incident by as much as 25% (if the demand could have been reduced by 10% in the August 12 incident) and saving as much as $2.4 million in the value of the motorists' time lost in an incident (if the demand could have been reduced by 10% in the September 7 incident). The most controversial approach is the third one, since even in the relatively cautious case considered here--where only the very farthest lane on the safer side of the highway is kept open during the incident--it might have led to an increase in risk in order to save waiting time. When (if ever) such tradeoffs are justified and how to evaluate them are questions that are beyond the scope of this paper, but as with other difficult social choices, there may very well be situations in which the price of extreme caution is too high and some risk must be accepted.

References

American Automobile Association. 1989. Potomac News (Falls Church, VA), April.

Deacon, R.T. and J. Sonstelie. 1985. "Rationing by Waiting and the Value of Time: Results from a Natural Experiment," J. of Political Economy vol. 93, no. 41, 627-647.

Fisher, A., L. Chestnut and D. Violette. 1989. "The Value of Reducing Risks of Death: A Note on New Evidence," J. Policy Analysis and Management vol. 8, 88-100.

Larson, R.C. 1988. "There's More to a Line than Its Wait," Technology Review, July, 60-67.

Metropolitan Washington Council of Governments. 1986. Solving the Problem of Greater Washington's Main Street (Washington, D.C.).

Morales, J.M. 1986. "Analytical Procedures for Estimating Freeway Traffic Congestion," Public Roads vol. 50, no. 2, 55-61.

Teal, R.F. 1988. Estimating the Full Economic Costs of Truck Incidents on Urban Freeways (Institute of Transportation Studies, University of California, Irvine UCI-ITS-RR-88-3), November.

Witzig, W.F. and J.K. Shilleen. 1987. Evaluation of Protective Action Risks (Washington, D.C., U.S. Nuclear Regulatory Commission NUREG/CR-4726), June.

A COMMUNITY-FOCUSED ROUTING AND SITING MODEL
FOR
HAZARDOUS MATERIALS AND WASTES

By George List[1], Member, ASCE,
and Pitu Mirchandani[2]

Abstract: A model is presented that shippers, carriers and policy-makers can use to analyze routing problems for hazardous materials or routing-and-siting problems for wastes. Unlike many other models, which focus on either the network or planar aspects of the problem, this one focuses on both. It ties the transportation costs to the network elements while the risk impacts relate to the zones. Routing and siting decisions produce impacts that accumulate, overlap and interact at the zonal level.

INTRODUCTION AND LITERATURE REVIEW

This paper presents a model that shippers, carriers and policy-makers can use to analyze routing and siting issues for hazardous materials and wastes. Its framework allows consideration of the tradeoffs among total risk, total cost, and the equitable distribution of risk. While transportation costs are tied to travel distances and times across the network, the risks relate to impacts measured directly at the zonal level.

The routing problem involves selecting "optimal" paths for hazardous material or waste shipments between one or more origin-destination (OD) pairs (e.g, "optimal" may mean a balance among

---

[1] Associate Professor, Department of Civil Engineering, Rensselaer Polytechnic Institute, Troy, NY 12180. (518)276-6362
[2] Professor, Department of Systems and Industrial Engineering, The University of Arizona, Tucson, AZ 85721. (602)621-2990

risk, cost, and the equitable distribution of risk). One such model (Batta and Chiu 1988) minimizes the number of people within "p-neighborhoods" of the links traversed by the route. Another (Robbins 1981) minimizes a linear combination of population exposure and shipment distance. A third (Saccomanno and Chan 1985) focuses on minimizing total risk, accident likelihoods, and truck operating costs. A fourth (Cox 1984) includes considerations of scheduling as well as routing as does its more recent extension (Turnquist and Cox 1986). A fifth model (Zografos and Davis 1986), uses goal programming to minimize (1) the total people exposed, (2) the affected people in special population categories such as school children, (3) travel cost, and (4) property damage. A sixth (Abkowitz and Cheng 1988) finds the convex hull of all non-dominated routes, minimizing linear combinations of risk and cost. The model presented here finds flow patterns which provide network-wide, non-dominated combinations of cost, risk, and risk equity.

Siting

The siting problem involves "optimally" locating a chain of treatment and disposal facilities (and choosing technologies). "Optimal" may again mean striking a balance among risk, cost and risk equity. One such model (Jennings and Sholar 1984) minimizes system cost, including treatment, disposal and transportation. Penalty functions account for the risk impacts involved. Four constraints are imposed: (1) all waste must be treated, (2) residues from these treatment processes must also be treated, (3) treatment capacity must not be exceeded at any disposal site for any treatment process and (4) disposal site capacities must not be exceeded. A second model (Mirchandani 1987) also minimizes total system cost. Its first component, operating cost, is defined as transportation plus treatment cost. The second, capital cost, is defined as the cost of building a treatment facility of a given type at a given location. More general models for siting "obnoxious facilities can also be found (e.g., Church and Garfinkel 1978, Dresner and Wesolowsky 1985).

Combined Routing and Siting

The routing and siting problem involves making these "optimal" decisions simultaneously: routing for the material shipments, and routing and siting for the wastes. The need for simultanaeity comes about because selection of a site does imply, implicitly, the selection of routes as well. Suitable paths must exist for the flows to take place.

Only a few routing and siting models exist. One (Shobrys 1981) minimizes a linear combination of ton-miles and "person-tons" of exposure. It also ensures that (1) each waste generator ships all it produces, (2) transshipment nodes forward all they receive, and

(3) storage facilities do not exceed their capacities. Another (Zografos and Samara 1989) minimizes preemptive goal deviations for (1) transportation risk, (2) travel time, and (3) disposal risk. Three constraints are involved: (1) all waste must be shipped, (2) a facility must be open if it is used and (3) only a limited number of facilities can be open.

MODEL FORMULATION

The model presented here assumes that a region consists of a set of zones on which a network of links, nodes, and potential waste treatment-disposal sites is superimposed.

It also assumes that each zone z experiences impacts from the shipments passing by and the processing taking place at nearby facilities. This is illustrated in Figure 1. The overlapping impacts result in distributions of "danger exposure" for each zone (see Figure 2). The distributions depend on the shipment patterns, the facility locations, the technologies employed, and other factors (e.g., environmental conditions). Such distributions result in a certain level of risk being perceived in each zone z.

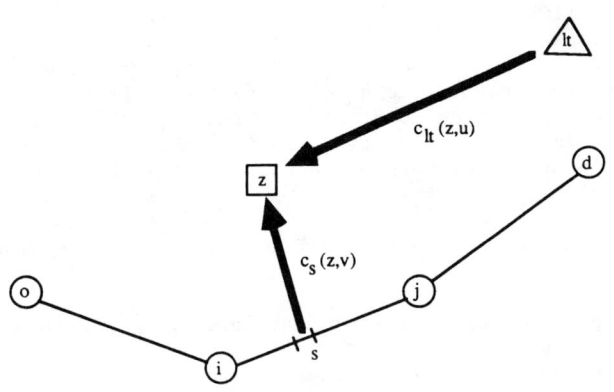

Schematic to indicate effects (bold arrows) of an incident on link segment s due to transporting volume v from node o to node d, and from treating volume u by technology t at site i.

FIG. 1. Route and Site Risk Impacts on Zones

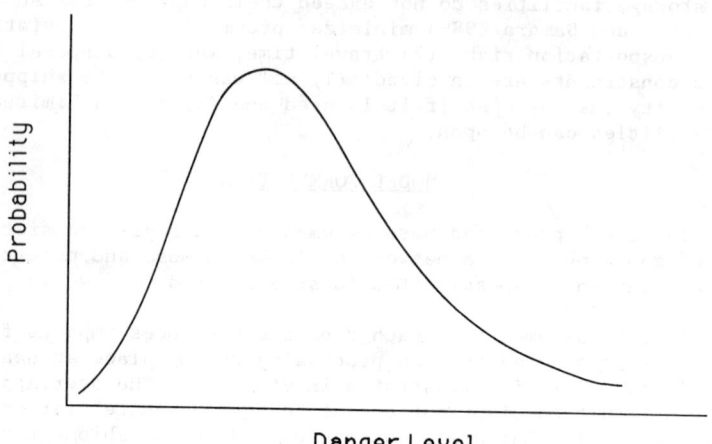

FIG. 2. Danger Distribution for a Zone

Mathematically, three functions are needed to make this model functional. The first is $\underline{c}_p(z,v_{pm})$, which provides the danger level distributions for zone z due to volume v of material type m passing over path p. The second is $\underline{c}_{lt}(z,q_w)$, which provides similar distributions for volume q_w of waste type w being processed by treatment technology t at location l. Finally, the third function, RISK($\underline{c}_p(z,v_{pm})$), converts these into the risk perceived by the people living and/or working in zone z. All three functions may be (and likely are) nonlinear.

To proceed further, we need to define the following variables and set names. Let:

P = set of paths (p∈P) for transporting materials or wastes
A = set of arcs (a∈A) in the network
N = set of nodes (n∈N) in the network
Z = set of zones (z∈Z) that comprise the geography of the plane
$\underline{c}_z$ = the vector of danger level distributions for zone z
r_z = perceived risk level for zone z

y_{max} = maximum risk exposure level per unit population among all zones
POP_z = the population in zone z.
$\$_p$ = the cost of using path p
M = set of materials (m∈M)
W = set of wastes (w∈W)
v_{pm} = the volume of material m using path p
L = set of sites (l∈L) where a treatment facility can be located.
$\$_l$ = the cost of building (opening) facility l
T = set of treatment technologies (l∈L) available.
O_w = set of nodes where wastes streams originate
P_{ol}^w = set of paths outbound from o∈O_w
v_{ol}^p = quantity of waste being transported on path p∈P_{ol}
δ_{lt}^{pw} = decision variable indicating whether treatment technology t is established at location l (δ_{lt} = 1) or not (δ_{lt} = 0)
θ_{pwt} = the fraction of v_{pw} treated by technology t
q_{ow}^{pwt} = the total waste (w∈W) generated at o∈O_w.

We will also restate definitions for the three main functions employed:

$\underline{c}_p(z, v_{pm})$ = function which provides the danger level distributions expected for zone z due to shipping volume v of material type m over path p

$\underline{c}_{lt}(z, q_w)$ = function which provides the danger level distributions for zone z due to volume q_w of waste type w being processed by treatment technology t at location l

$RISK(\underline{c}_p(z, v_{pm}))$ = function which translates the danger level distributions into a perceived risk level for zone z

The model itself is given by the following set of equations which form a non-linear, multi-objective programming problem:

<u>Minimize</u>:

o Total cost (transportation, treatment and facility):

$$TC = \sum_p \sum_{k \in M+W} \$_p v_{pk} + \sum_l \sum_t \$_{lt} \delta_{lt} \qquad (1a)$$

o Total risk

$$TR = \sum_z r_z \qquad (1b)$$

o Maximum risk per unit population (equity)

$$EQ = y_{max} \qquad (1c)$$

<u>subject to</u>:

o computing the risk for each zone z

r_z = (RISK ($\underline{c}_z$) POP$_z$ for all $z \in Z$ \hfill (1d)

   o identifying the largest of these risks

$y_{max} \geq$ RISK ($\underline{c}_z$) for all $z \in Z$ \hfill (1e)

   o ensuring that all material is shipped:

$\sum_{p \in P_{od}} v_{pm} = V_{odm}$ for all $(o,d) \in O,D$, $m \in M$ \hfill (1f)

   o all waste on a given path is treated:

$\sum_t \Theta_{pwt} = 1$ for all $p \in P_{ol}$, $o \in O_w$, $w \in W$ \hfill (1g)

   o treating waste by technology t at site l implies $\delta_{lt} > 0$:

$\Theta_{pwt} \leq \delta_{lt}$ for all $l \in L$, $t \in T$, $p \in P_{ol}$, $o \in O_w$, $w \in W$ \hfill (1h)

   o all wastes are routed to some treatment facility:

$\sum_l \sum_{p \in P_{ol}} v_{pw} = q_{ow}$ for all $o \in O_w$ and $w \in W$ \hfill (1i)

   o the total inbound waste is the sum of the inbound flows

$u_{ltw} = \sum_{o \in O_w} \sum_{p \in P_{ol}} \Theta_{pwt} v_{pw}$ for all $l \in L$, $t \in T$, and $w \in W$ \hfill (1j)

EXAMPLE APPLICATION

Hypothetical Network

Figure 3 shows a hypothetical urban area that has been used to test the model. It contains both radial and circumferential freeways. Three sites for building treatment facilities are available. Two are local: one in the northeast and one in the southwest. The third lies on the eastern periphery, at a distance whose value is allowed to vary. Fifteen links are in the network; 14 lie within the region, and the fifteenth leads to site S3.

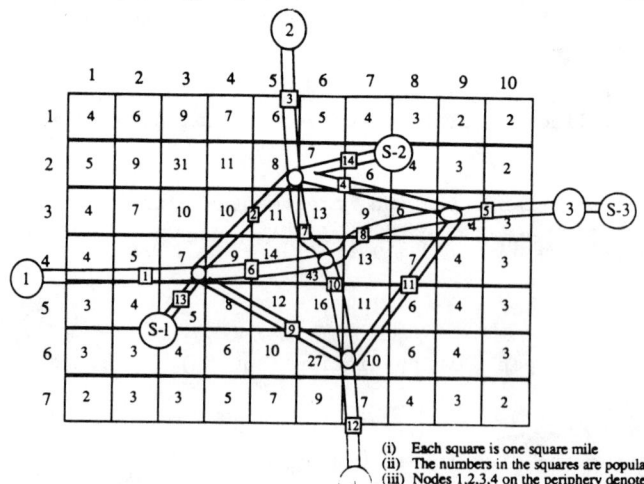

FIG 3. Hypothetical Urban Area

All hazardous material shipments originate and terminate at the boundary nodes (1, 2, 3, or 4). The waste shipments originate at these same nodes and are shipped to either node S1, S2, or S3 for treatment and disposal.

A Routing Only Scenario

We begin by discussing results for a scenario where only material shipments exist (We will call this problem "M").

First, as the weights on the three objectives vary, only seven non-dominated flow patterns emerge. This suggests that these solutions are quite robust. Even though the decisionmaker might be unsure of the exact weight values, as long as they lie within certain ranges, he or she can be confident that the solution found will be the best one to select.

Another observation is that the flow patterns have many features in common. For example, OD pairs that are adjacent around the circumference of the region (e.g., 2-3) always use the beltway because they offer the lowest possible cost and risk. In addition, both the minimum risk and minimum cost solutions use only one route for each OD pair. This is because both objectives are linear (i.e., $\Sigma c_{ij} x_{ij}$) and no capacity constraints are imposed on the links. However, for the OD pairs directly across from one another (i.e., 1-3, 3-1, 2-4, and 4-2) the routes change between the solutions. For minimum risk, beltway-based routes are used whereas for minimum cost, those through the center of the region are employed. Then, for the most equitable risk distribution, the beltway-based routings are again used. However, for two OD pairs (3-1 and 4-2) their flows are split between the two beltway routes available (e.g., for 3-1, the one which is northern and counter-clockwise and the other which is southern and clockwise).

As Figure 4 shows, the tradeoffs among the objectives are noticable but minor, with 10% to 20% penalties being involved among the objectives. The risk penalty for achieving equity, however, appears to be greater than the converse, suggesting that minimizing total risk at the expense of equity may not be a desirable strategy. Also, since the cost penalty involved is never very great (less than 20% and often 10 or 15%, cost should probably not be a major factor to consider). By way of contrast, the risk and equity penalties involved in minimizing cost are very large, being 42% and 64% respectively; this is not shown in the figure.

Combined Routing and Siting Scenarios

Three combined routing and siting scenarios are presented. We will refer to them as problems MW1, MW2 and MW3. The difference among them lies in the assumed distance, D_{15}, to external treatment site S3. In problem MW1, D_{15} is 1.5 miles, in MW2 it is 0.5 miles, and in MW3 it is -0.5 miles (i.e., 0.5 miles inside the regional boundary.)

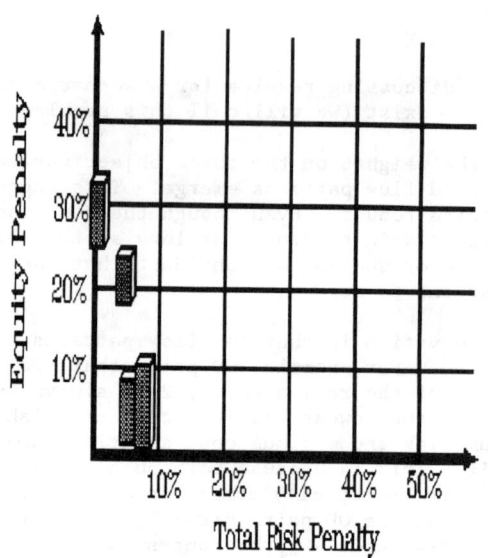

FIG. 4. Objective Tradeoffs for the Material Routing Scenario
Bar heights denote cost penalties. Those shown are 10, 15 and 20%

Problem MW1. Problem MW1 comes close to being like problem M. Treatment site S3 is effectively far enough away that its processing risks have no effect on the region (it is not in the region's back yard).

Several parallels can be drawn between problem MW1 and problem M. First, their minimum cost flow patterns are nearly identical. Second, the same is true for the minimum risk solutions. Third, in problem MW1, nine of the 12 minimum cost routes are identical to those for the minimum risk solution (it was 10 out of 12 for problem M).

Another observation is that the waste is almost always shipped to S3 for treatment. (In fact, given the assumed parameters for the impacts, D_{15}=1.5 miles appears to be the distance at which such impacts become zero. We found in test cases involving values of D_{15} greater than 1.5, that the optimal flow patterns matched those for problem MW1.) Only in the minimum total cost solution and in two others having a high weight on equity, is waste shipped to a local treatment site. Even then, the amount is small (6% of the total shipments).

The equity solution is quite different. The model elects to ship the wastes from origin 2 to S2. This implies that the marginal gain in equity from treating source 2 waste at S2 is greater than that from shipping it to S3. For real world situations, it means that there might be times when it is more advantageous to treat the wastes locally than to ship them to a distant treatment site.

Insofar as tradeoffs among the objectives are concerned, Figure 5 shows that the equity and risk penalties parallel those found for problem M as might be expected since the treatment occurs predominantly at S3 and the added volume involved due to the wastes is small (10% more than in problem M).

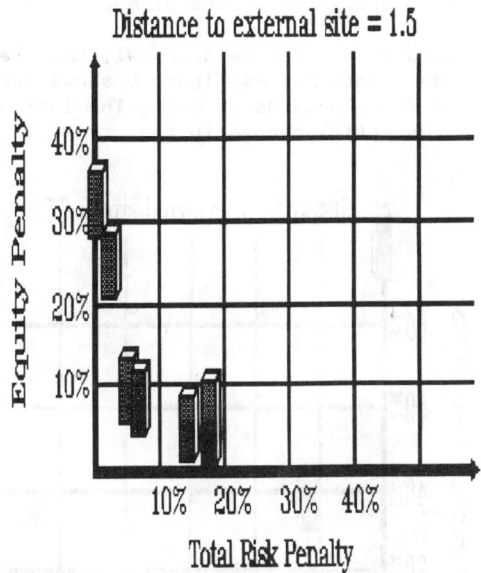

FIG. 5. Objective Tradeoffs for Routing and Siting Scenario MW1
Bar heights denote cost penalties. Those shown are 15 and 20%

We also see that the burden of making optimal tradeoffs among the objectives rests on the routing rather than siting decisions. This is true, of course, because most of the treatment occurs at S3. In all but the minimum risk and cost solutions, at least one OD pair has traffic assigned to two routes. Vis a vis problem M, the volume of material and waste being shipped is, in total, at a level where partial assignments are key to the equity objective almost regardless of the weights applied to the objectives.

Finally, the "through-the-center" links are used only in the optimal equity and minimum total cost solutions. Otherwise, the beltway links are used. Apparently, the equity objective can be satisfied without shipments through the central area, a condition that changes radically as S3 is moved closer to the region.

<u>Problem MW2.</u> In problem MW2, D_{15} is 0.5 miles. The routing and siting patterns here are markedly different from those in MW1. First, the burden of making optimal tradeoffs among the objectives rests on the waste-related decisions, especially siting. All the material flows use single routes regardless of the weights. In contrast, partial shipments of the wastes are common: among treatment sites for a given origin, and among routes for a given OD pair. For example, the wastes from origins 1 and 4 are typically split between sites S1 and S2 (with one or more routes to each).

Second, in contrast to problems M or MW1, the tradeoffs among the objectives are substantial. As Figure 6 shows, equity and total risk penalties reach upwards of 80%. The introduction of siting has made a substantial change in the tradeoffs involved.

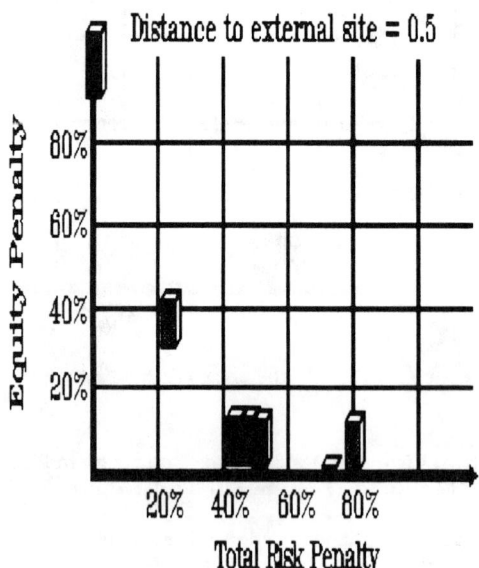

FIG. 6. Objective Tradeoffs for Routing and Siting Scenario MW2
Bar heights denote cost penalties. Those shown are 0, 10 and 15%

Third, S3 is always used (because its impacts are still the lowest of the three sites). And S2 is always given a heavier processing load than S1. (This means that S3 should be larger than S2; and S2, larger than S1, regardless of the objective weights.)

Last, shipments use the "through-the-center" links more often than in MW1. They are routed this way to achieve an equitable distribution of risk. The "high" local risk impacts felt by the zones near S1 and S2 are balanced by shipments through the central urban area. (There may be a dilemma here.)

<u>Problem MW3</u>. Problem MW3 makes the trends identified in MW2 even clearer. Material flows again use single routes, regardless of the weights employed. The wastes are "carefully distributed" among S1, S2, and S3 for treatment. Partial assignments are even more common; to treatment sites, and among routes for a given origin-treatment site pair. The use of "through-the-center" links also becomes more common. In addition, as Figure 7 shows, the tradeoffs among the objectives are even more pronounced. The equity penalty required to achieve minimum risk involves a 114% increase in the equity objective function.

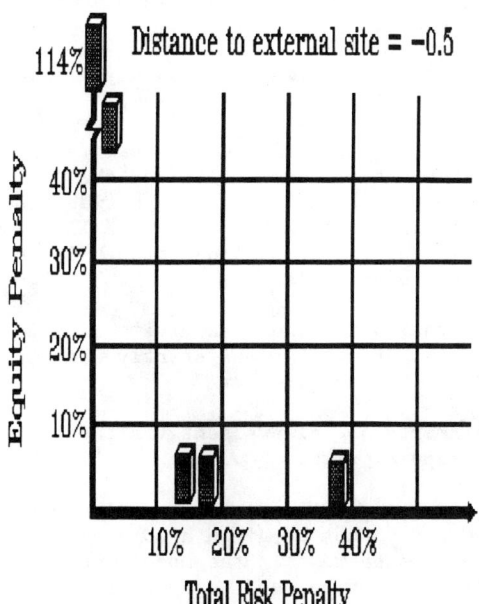

FIG. 7. Objective Tradeoffs for Routing and Siting Scenario MW3

Insights

From these case studies we conclude the following. First, when waste shipments are involved, they should be routed carefully to achieve an optimal balance among the objectives.

Second, the more the waste treatment is localized, the more the radial freeway links need to be used to achieve an equitable distribution of risk. In essence, placing the treatment sites in the low density areas raises the risk exposure there and creates an "inequitable" distribution of risk. The corrective action taken is to increase the risk exposure in the high density areas by routing some shipments through them. There may be a dilemma here.

We should also observe that the model's preference for S2, vis a vis S1, pertains throughout the MW problems. Therefore, S2 should be sized larger than S1 if the two facilities are built.

What Else do the Results Indicate?

Several additional conclusions can be drawn based on the case studies conducted:

o Significant tradeoffs do exist among equity, total risk and total cost.

o Pursuing a strict minimum risk solution produces very inequitable distributions of risk.

o As long as equity is weighted at least as heavily as risk or cost, the traffic passing *across* the metropolitan area (i.e., 1-3, 3-1, 2-4, 4-2) splits between routes that use the beltway links (links 2, 4, 9, and 11 as appropriate). In general, the routes passing through the heart of the area are not used.

o Shipments that skirt the area (i.e., 1-2, 2-1, 2-3, 3-2, 3-4, 4-3, 4-1, and 1-4) consistently uses the beltway (links 2, 4, 9, and 11).

o Waste shipments are always sent via the beltway to the nearest disposal site.

Of course, these conclusions pertain only to the example problems analyzed and cannot be given a more general interpretation. It is clear, however, that the model, when applied to a given network setting, can be used to develop such insights and conclusions. At present, work is underway to apply the methodology to a real-world problem.

SUMMARY AND RECOMMENDATIONS

This paper has presented an integrated model that decision-makers from both industry and government can use to analyze routing and siting problems for hazardous materials and wastes. The model explicitly considers both the network on which the routes and sites are selected and the underlying plane (zones) where the risk impacts are felt. The model can enable industry to optimize its logistics plans for given a regulatory environment. It also can allow governmental agencies to determine how regulations affect routing and siting patterns.

ACKNOWLEDGEMENTS

The authors would like thank S. Krishnamoorthy for programming the model and helping us get the computational results for the hypothetical problem. We would also like to express their appreciation to for the suggestions provided by S. Raghavan, R. Rebello, and G. Beroggi who reviewed early drafts of the paper.

REFERENCES

Abkowitz, M. and Cheng, P. (1988). "Developing a Risk-Cost Framework for Routing Truck Movements of Hazardous Materials." Accident Analysis and Prevention, 39-51.

Batta, R. and Chiu, S. (1988). "Optimal Obnoxious Paths on a Network: Transportation of Hazardous Materials." Operations Research, 36, 84-92.

Church, R., and Garfinkel, R. (1978). "Locating an Obnoxious Facility on a Network." Transportation Science, 12, 107-118.

Cox, R. (1984). "Routing and Scheduling of Hazardous Materials Shipments: Algorithmic Approaches to Managing Spent Nuclear Fuel Transport," thesis presented to Cornell University, at Ithaca, New York, in partial fulfillment of the requirements for the degree of Doctor of Philosophy.

Drezner, A. and Wesolowsky, G. (1985). "Location of Multiple Obnoxious Facilities." Transportation Science, 19, 193-202.

Jennings, A., and Sholar, R. (1984). "Hazardous Waste Disposal Network Analysis." Journal of Environmental Engineering, ASCE, 110, 325-342.

Mirchandani, P. (1987). "A Model-Based Approach to Planning Industrial Toxic Waste Sites." Hazardous Materials Disposal: Siting and Management, (ed. Chjatterji, M.), Gower Publishing Company.

Robbins, J. (1981). "Routing Hazardous Materials Shipments," thesis presented to Indiana University, at Bloomington, Indiana, in partial fulfillment of the requirements for the degree of Doctor of Philosophy.

Saccomanno, F. and Chan, A. (1985). "Economic Evaluation of Routing Strategies for Hazardous Road Shipments." Transportation Research Record 1020, 12-18.

Shobrys, D. (1981). "A Model for the Selection of Shipping Routes and Storage Locations for a Hazardous Substance," thesis presented to John Hopkins University, at Baltimore, Maryland, in partial fulfillment of the requirements for the degree of Doctor of Philosophy.

Turnquist, M. and Cox., R. (1986). "Scheduling Truck Shipments of Hazardous Materials in the Presence of Curfews." Transportation Research Record 1063, 21-26.

Zografos, K., and Davis, C. (1986). "A Multiobjective Model for Routing Hazardous Materials: A Goal Programming Approach." Technology Transfer Center, University of Connecticut, Storrs, Connecticut.

Zografos, K. and Samara, S. (1989). "A Combined Location-Routing Model for Hazardous Waste Transportation and Disposal." Department of Civil and Architectural Engineering, University of Miami, Coral Gables, Florida.

Truck Accident and Release Rates for Hazmat Routing

Douglas W. Harwood[1]
John G. Viner[2]
Eugene R. Russell[3]

Abstract

Estimates of accident and release rates are essential in the conduct of risk assessments in routing studies for highway transportation of hazardous materials. Recent published literature underscores both the importance of these rates in risk assessment and the significant shortcomings of the available data. Truck accident rates are developed in this paper as a function of roadway type and area type (urban/rural) from state data on highway geometrics, traffic volume, and accidents. Release probabilities in accidents are derived from a combination of federal and state truck accident data bases. A revised model for the accident probability portion of the DOT hazardous materials routing guidelines is recommended. Statistical tests are presented to determine whether accident rates based on site-specific data or systemwide values, such as those derived in this paper, should be used for any particular route segment.

Introduction

The most widely accepted risk assessment model for identifying preferred routes for hazardous materials transportation is that presented in the DOT guidelines. This model was first presented in the 1980 FHWA publication, "Guidelines for Applying Criteria to Designate Routes for Transporting Hazardous Materials" (Barber and Hildebrand, 1980). This document was recently updated and republished by the DOT Research and Special Programs Administration (Research and Special Programs Administration, 1989).

---

[1]Midwest Research Institute, 425 Volker Boulevard, Kansas City, MO 64110; [2]Federal Highway Administration, 6300 Georgetown Pike, McLean, VA 22101; [3]Kansas State University, Civil Engineering Department, Manhattan, KS 66506.

The DOT guidelines are based on selection of minimum risk routes, where risk is determined for individual route segments as:

Risk = Accident Probability x Accident Consequences   (1)

The DOT guidelines contain procedures for determining accident risk on the basis of accident rate and route segment length and procedures for determining accident consequences on the basis of either the number of persons potentially exposed or the value of property potentially exposed to hazardous materials releases. This paper focuses on updated procedures and improved data for assessing the accident probability term in Equation (1).

A recent critique of the DOT guidelines has identified several potential approaches to strengthening the accident probability portion of the guidelines (Harwood and Russell, 1989). These recommendations are based on several perceived weaknesses in the current guidelines. These include:

- The default values of accident rate in the DOT guidelines are based on accident predictive models that are 15 to 20 years old, and may be out of date (Smith, 1973; Mulinazzi and Michael, 1967; and Urbanek and Barber, 1980).

- These models apply to accident rates for all vehicle types rather than to truck accident rates. All-vehicle accident rates are based primarily on passenger car accidents, while highway transportation of hazardous materials is conducted by truck.

- The DOT guidelines implicitly assume that all accidents are equally likely to result in a hazardous materials release. In fact, recent research has established that some types of accidents are much more likely than others to result in a release (Harwood and Russell, 1989; and Harwood, Russell, and Viner, 1989).

- The guidelines recommend the use of observed accident rates for the specific route segments under analysis in preference to the default values whenever possible. However, no statistical guidance is given on whether the observed accident rate is based on a sufficiently large sample of accidents to be statistically reliable and whether the differences between the observed accident rates and the default values are statistically significant.

A recently published paper by Glickman and its accompanying discussion by Abkowitz clearly illustrate the significant shortcomings in quality in much of accident rate data currently available for hazardous materials transportation risk assessments (Glickman, 1988). This paper addresses the need for better data through the development from existing federal and state data bases of truck accident rates and hazardous materials release probabilities that can be used in lieu of the default values presented in the DOT guidelines.

Determination of Truck Accident Rates

A key element in comparing the risks of alternative routes for hazardous materials transportation is having reliable data on truck accident rates for use in determining the relative probabilities of hazardous materials releases. An important difference between alternative routes that must be accounted for in routing studies is the difference in truck accident rates between roadway and area types. For example, freeways generally have lower accident rates than other types of highways and urban highways (especially nonfreeways) generally have higher accident rates than rural highways. These differences between highway and area types are well known for all-vehicle accident rates, but have only been demonstrated for trucks in studies based on a limited number of highway sections (Harwood and Russell, 1989; Graf and Archuleta, 1985; and Carsten, 1987). Therefore, truck accident rates for use as default values in hazardous materials routing were developed from existing data bases containing highway geometrics, truck volume, and truck accident for the entire state highway system of three states: California, Illinois, and Michigan. These states were chosen because they were found to have the most complete, mergeable, and easy-to-use computer files.

Truck accident rates for each highway class of interest we computed as:

$$TAR_j = \sum_i \frac{A_{ij}}{VMT_{ij}} \qquad (2)$$

where: TAR_j = Average truck accident rate for highway class j

A_{ij} = Number of accidents in one year on route segment i in highway class j

VMT_{ij} = Annual vehicle-miles of travel on route segment i in highway class j

The procedure described above was applied to existing geometric, traffic volume, and accident files for the entire state highway systems of California, Illinois, and Michigan that could be linked by mileposts. Tables 1 and 2 show the determination of truck accident rates and truck accident type distributions, respectively, for California. Similar tables were also prepared for Illinois and Michigan state highways. Table 3 shows the average truck accident rates for each highway class in each state and the weighted three-state average.

The truck accident rates given in Table 3 are appropriate for use as default values in hazardous materials routing studies where data more suited to local conditions are not available. Highway agencies are encouraged to develop comparable default values from their own data, wherever possible.

Determination of Hazmat Release Probabilities

The probability portion of the DOT routing guidelines is based entirely on accident probabilities. Of course, an accident involving a hazmat-carrying truck cannot lead to potentially catastrophic consequences unless the hazardous materials being transported is released. Thus, the current risk assessment methodology implicitly assumes that hazmat releases are equally likely in all accidents.

A recent FHWA study has shown that the probability of a hazmat release given an accident involving a hazmat carrying truck varies markedly with the type of accident (Harwood and Russell, 1989). Table 4, based on data from the FHWA motor carrier accident reports, shows that release probabilities are highest in single-vehicle noncollision accidents and truck-train collisions and lowest in multiple vehicle collisions. Furthermore, it is known that the various highway classes have distinctly different patterns of accident types. For example, the percentage of single-vehicle noncollision accidents (which have the highest probability of producing a hazmat release if an accident occurs) is about twice as high on rural highways as on urban highways. Therefore, the probability portion of the DOT guidelines should include a term representing the probability of release given an accident. Default values for this term are developed below.

Table 1. Truck accident rates on California state highways, 1985-1987.

Highway class Area type	Roadway type	Total length (mi)	No. of sections	Average truck ADT (veh/day)	No. of truck accident involvements[a]	Truck travel (MVM)	Truck accident rate[b] (per MVM)
Rural	Two-lane	8,808.96	2,607	392	6,577	3,784.97	1.73
Rural	Multilane undivided	209.13	334	858	1,070	196.58	5.44
Rural	Multilane divided	726.85	450	1,839	1,801	1,463.45	1.23
Rural	Freeway	2,068.20	405	4,791	5,759	10,850.90	0.53
Rural	TOTAL	11,813.14	3,796	1,260	15,207	16,295.90	0.93
Urban	Two-lane	513.49	648	748	1,778	420.69	4.23
Urban	Multilane undivided	141.50	341	1,116	2,251	172.84	13.02
Urban	Multilane divided	754.18	793	1,644	4,996	1,427.47	3.50
Urban	One-way street	22.26	47	1,387	223	33.81	6.60
Urban	Freeway	1,969.65	817	8,395	28,860	18,107.00	1.59
Urban	TOTAL	3,401.07	2,646	5,414	38,108	20,161.81	1.89
TOTAL		15,214.21	6,442	2,388	53,315	39,781.10	1.34

[a] Accidents involving two or more trucks are counted as two or more involvements.
[b] Computed from Equation (4).

182 HAZARDOUS MATERIALS TRANSPORTATION

Table 2. Truck accident type distribution on California state highways, 1985-1987.

		Percent of accident involvements								
		Single-vehicle noncollision accidents			Single-vehicle collision accidents				Multiple-vehicle collision accidents		
Area type	Highway class Roadway type	Run-off road	Overturned	Other	Coll. w/ parked vehicle	Coll. w/ train	Coll. w/ nonmotorist[a]	Coll. w/ fixed object	Other collision	Collision w/passenger car	Coll. w/truck	Coll. w/other vehicle
Rural	Two-lane	4.5	6.6	4.4	2.4	0.0	0.6	7.0	5.7	29.8	26.6	12.4
Rural	Multilane undivided	3.6	7.5	3.9	4.3	0.0	0.4	7.5	5.7	27.4	26.1	13.7
Rural	Multilane divided	3.6	4.0	3.8	3.9	0.0	0.2	6.1	4.7	33.4	26.4	13.8
Rural	Freeway	3.5	3.3	3.8	3.8	0.0	0.4	7.4	5.0	31.3	22.3	19.4
Rural	TOTAL	3.9	5.1	4.1	3.2	0.0	0.5	7.1	5.3	30.6	24.9	15.3
Urban	Two-lane	1.5	2.6	3.4	3.6	0.0	0.3	5.1	3.9	39.6	30.7	9.3
Urban	Multilane undivided	0.2	0.6	2.6	8.5	0.0	0.8	5.1	4.0	41.3	30.1	6.9
Urban	Multilane divided	0.8	1.3	2.4	7.0	0.0	0.6	5.7	3.8	43.7	28.1	6.6
Urban	One-way street	0.0	2.2	0.9	9.4	0.0	1.3	6.3	2.2	45.7	27.4	4.5
Urban	Freeway	0.6	1.0	1.3	1.9	0.0	0.2	3.2	1.7	50.6	25.6	13.9
Urban	TOTAL	0.6	1.1	1.6	3.1	0.0	0.3	3.8	2.2	48.6	26.4	12.3
TOTAL		1.6	2.3	2.3	3.1	0.0	0.4	4.7	3.1	43.4	26.0	13.1

[a] Nonmotorists include animals, pedestrians, and bicycles.

Table 3. Truck accident rates by state and combined.

Highway class		Truck accident rate (accidents per million veh-mi)			
Area type	Roadway type	California	Illinois	Michigan	Weighted average[a]
Rural	Two-lane	1.73	3.13	2.22	2.19
Rural	Multilane undivided	5.44	2.13	9.50	4.49
Rural	Multilane divided	1.23	4.80	5.66	2.15
Rural	Freeway	0.53	0.46	1.18	0.64
Urban	Two-lane	4.23	11.10	10.93	8.66
Urban	Multilane undivided	13.02	17.05	10.37	13.92
Urban	Multilane divided	3.50	14.80	10.60	12.47
Urban	One-way street	6.60	26.36	8.08	9.70
Urban	Freeway	1.59	5.82	2.80	2.18

[a] Weighted by veh-mi of truck travel.

Table 4. Probability of release given that an accident has occurred as a function of accident type.

Accident type	Probability of release
SINGLE-VEHICLE NONCOLLISION ACCIDENTS	
Run-off-road	0.331
Overturned (in road)	0.375
Other noncollision	0.169
SINGLE-VEHICLE COLLISION ACCIDENTS	
Collision with parked vehicle	0.031
Collision with train	0.455
Collision with nonmotorist	0.015
Collision with fixed object	0.012
Other collision	0.059
MULTIPLE-VEHICLE COLLISION ACCIDENTS	
Collision with passenger car	0.035
Collision with truck	0.094
Collision with other vehicle	0.037

Table 4 was developed from the FHWA motor carrier accident reports because, for each accident involved truck, this data base documents both whether the truck was carrying hazardous materials and whether the hazardous materials were released. It would be desirable for users to derive values comparable to those in Table 4 for their own state, but only three states (Louisiana, Missouri, and Wyoming) currently have both data items needed to make this determination in their accident records systems (Harwood and Russell, 1989).

The probability of a hazmat release given an accident varies between highway classes because it varies with accident type and because the distribution of accident types varies markedly between highway classes. For example, Table 2 shows that the proportion of single-vehicle noncollision accidents (which are very likely to result in a hazmat release) is nearly 50 percent higher on rural two-lane highways than it is on rural freeways. The probability of a release given an accident involving a hazmat-carrying vehicle for a particular highway class can be computed as:

$$P(R|A)_j = \sum_k P(R|A)_k \times P(k)_j \tag{3}$$

where: $P(R|A)_j$ = Probability of a hazmat release given an accident involving a hazmat carrying vehicle for highway class j

$P(R|A)_k$ = Probability of a hazmat release given an accident involving a hazmat carrying vehicle for accident type k (from Table 4 or equivalent state data)

$P(k)_j$ = Probability that an accident on highway class j will be of accident type k (i.e., proportion of truck accidents for each accident type shown in Table 2 on highway class j from state accident data)

The probabilities in Table 5 are appropriate for use as default values in hazmat routing studies if data more suited to local conditions are not available.

Table 5. Probability of hazmat release given that an accident has occurred as a function of highway class.

Highway class		Probability of hazmat release given an accident			
Area type	Roadway type	California	Illinois	Michigan	Weighted average[a]
Rural	Two-lane	0.100	0.074	0.073	0.086
Rural	Multilane undivided	0.100	0.071	0.064	0.081
Rural	Multilane divided	0.087	0.064	0.062	0.082
Rural	Freeway	0.083	0.111	0.095	0.090
Urban	Two-lane	0.077	0.059	0.069	0.069
Urban	Multilane undivided	0.064	0.052	0.055	0.055
Urban	Multilane divided	0.068	0.048	0.058	0.062
Urban	One-way street	0.066	0.050	0.056	0.056
Urban	Freeway	0.062	0.055	0.067	0.062

[a] Weighted by veh-mi of truck travel.

Revised Procedures for Determining Accident Probabilities

In the current DOT guidelines, the probability of a hazmat accident is computed in the risk assessment model from the following equation:

$$P(A)_i = AR_i \times L_i \qquad (4)$$

where: $P(A)_i$ = probability of a hazardous materials accident for route segment i

AR_i = accident rate per veh-mi for all vehicle types on route segment i

L_i = length (mi) for route segment i

The availability of the truck accident rate and release probabilities presented above permits estimation of the probability of a hazmat accident in which a release occurs. The probability of a releasing accident should be computed with the following equation, which could replace Equation (4) in the DOT guidelines:

$$P(R)_i = TAR_i \times P(R|A)_i \times L_i \qquad (5)$$

where: $P(R)_i$ = probability of an accident involving a hazmat release for route segment i

TAR_i = truck accident rate (accidents per veh-mi) for route route segment i

$P(R|A)_i$ = probability of a hazmat release given an accident involving a hazmat-carrying truck for route segment i

L_i = length (mi) of route segment i

Equation (5) is more appropriate for hazmat routing analyses than Equation (4) because (1) risk is based on the probability of a hazmat release rather than just the probability of an accident; and (2) risk is based on truck accident rates rather than all-vehicle accident rates. Equation (5) retains the proportionality of risk to route segment length which is central to all routing analyses.

Table 6 presents typical values of truck accident rate and release probability taken from Tables 3 and 5 that can be used as default values in Equation (7). However, users are encouraged to develop default values based on average data for their own jurisdiction. A key aspect of Table 6 is that both truck accident rates and release probabilities vary with area type (urban/rural) and roadway type.

The DOT guidelines encourage users to base accident rates on site-specific accident histories, whenever possible. The guidelines do not appear to recognize the need for caution in using accident rates based on small sample sizes of accidents, which are typical of the relatively short route segments often used in risk assessments. For example, consider three 0.5-mi route segments on alternative routes. Suppose that, in a 3-year period, one of these segments experiences zero truck accidents, another experiences one truck accident, and the third experiences two truck accidents. It would certainly be incorrect to treat the first segment as having zero risk of a hazmat release, but this is the conclusion one would reach using the site-specific accident rate in Equation (4).

Table 6. Default truck accident rates and release probability for use in hazmat routing analyses.

Area type	Roadway type	Truck accident rate (accidents per million veh-mi)	Probability of release given an accident	Releasing accident rate (releases per veh-mi)
Rural	Two-lane	2.19	0.086	0.19
Rural	Multilane undivided	4.49	0.081	0.36
Rural	Multilane divided	2.15	0.082	0.18
Rural	Freeway	0.64	0.090	0.06
Urban	Two-lane	8.66	0.069	0.60
Urban	Multilane undivided	13.92	0.055	0.77
Urban	Multilane divided	12.47	0.062	0.77
Urban	One-way street	9.70	0.056	0.54
Urban	Freeway	2.18	0.062	0.14

It would also be incorrect to presume that, since the third segment has twice as many accidents as the second segment, it also has twice as much risk. The guidelines could be revised to incorporate a minimum time period or a minimum number of accidents needed to establish reliable accident rates. However, since default values of accident rate are available, it is usually more appropriate to rely on default values of accident rate for specific highway classes (e.g., rural two-lane highways, urban freeways, etc.) developed on a statewide or systemwide basis. An exception to this general rule occurs when the accident frequency for a specific route segment is either substantially higher or lower than the systemwide accident rate for its highway class. Since accident occurrence is a random variable, site-specific accident data cannot be presumed to indicate true differences in risk between segments unless a statistical test indicates that these differences are statistically significant.

In most cases, the truck accident rates shown in Table 6 or, better yet, average values for the user's own jurisdiction should be used as the value of TAR_i in Equation (5). However, a simple statistical procedure, based on the Chi-squared test, can be used to determine whether the actual accident frequency for a particular route segment is enough larger or smaller than the expected accident frequency to warrant replacement of the default truck accident rates by site-specific rates based on accident histories. This test is performed by calculating the value of the Chi-squared statistic (χ^2) is employed as follows:

$$\chi^2 = \frac{(A_e - A_o)^2}{A_e} \qquad (6)$$

where:

χ^2 = Chi-squared statistic

A_e = expected number of truck accidents

A_o = observed number of truck accidents

If $\chi^2 \leq 4$, then the expected and observed number of accidents do not differ significantly at the 5% significance level. Therefore, the systemwide default accident rate should be preferred to site-specific accident data.

If $\chi^2 > 4$, then the expected and observed number of accidents differ significantly. This indicates at the 5% significance level that the observed accident rate is lower or higher than the systemwide default value. In this case, the systemwide default accident rate should be replaced by a value based on the site-specific data. If the site-specific accident rate is greater than the default accident rate, then use the site-specific rate. If the site-specific accident rate is less than 50% of the default accident rate, then use 50% of the default accident rate. The latter restriction is based on judgment and is included to keep very low short-term accident experience, or poor accident reporting levels in a particular jurisdiction, from causing misleading results. Even if the roadway segment has experienced no accidents during the study period, there is still risk involved in transporting hazardous materials over the segment, and the use of 50% of the default accident rate is recommended.

A more detailed discussion of the Chi-squared procedure and an alternative procedure based on the Poisson distribution for use whenever $A_e < 5$ is provided in the full report by Harwood and Russell (1989).

Conclusions

This paper has shown how the accident probability portion of the DOT hazmat routing guidelines can be realigned to more realistically address the likelihood of accidents involving hazmat releases. Equation (5) presents the recommended method for determining the relative probability of a hazmat release for hazmat shipments on a particular route segment. The key elements in the revised guidelines are explicit consideration of the truck accident rates and the probability of a release given an accident. Truck accident rates are more directly applicable to the risk of accidents involving hazmat-carrying vehicles than the all-vehicle accident rates used in the current FHWA guidelines. Furthermore, the inclusion of hazmat release probabilities which vary markedly between accident types makes the revised procedures more sensitive to differences in accident patterns between highway types (e.g., freeway vs. nonfreeway).

The revised procedures are equally applicable to routing decisions based on a highway agency's own truck accident data and decisions based on the default values of truck accident rate and release probability presented in this paper. The use of truck accident rates based on an agency's own data is generally preferable, because these values will be most suited to local conditions.

The paper also emphasizes that site-specific accident data must be used cautiously when the available accident sample sizes for a particular route segment are small, as they often area. The Chi-squared test has a key role in the decision to use either the default value of truck accident rate or the truck accident rate based on site-specific data for any given route segment. Finally, in the special case where the expected number of truck accidents is less than 5, a test based on the Poisson distribution should be used in place of the Chi-squared test.

Acknowledgements

The work reported in this paper was conducted under sponsorship of the Federal Highway Administration. However, the findings and conclusions in the paper are those of the authors and do not necessarily represent the views of the Federal Highway Administration.

References

Barber, E. J., and L. K. Hildebrand, "Guidelines for Applying Criteria to Designate Routes for Transporting Hazardous Materials," Report No. FHWA-IP-80-15, Federal Highway Administration, November 1980.

Carsten, O., "Safety Implications of Truck Configuration," presented at 66th Annual Meeting of the Transportation Research Board, January 1987.

Glickman, T. S., "Benchmark Estimates of Release Accident Rates in Hazardous Materials Transportation by Rail and Truck," *Transportation Research Record 1193*, 1988.

Graf, V. D., and K. Archuleta, "Truck Accidents by Classification," Report No. FHWA/CA/TE-85, California Department of Transportation, January 1985.

Harwood, D. W., and E. R. Russell, "Present Practices of Highway Transportation of Hazardous Materials," Report No. FHWA-RD-89-013, Federal Highway Administration, November 1989.

Harwood, D. W., E. R. Russell, and J. G. Viner, "Characteristics of Accidents and Incidents in Highway Transportation of Hazardous Materials," presented at the 68th Annual Meeting of the Transportation Research Board, January 1989.

Mulinazzi, T. E., and H. L. Michael, "Correlation of Design Characteristics and Operational Controls with Accident Rates on Urban Arterials," Proceedings of the 53rd Annual Road School, *Engineering Bulletin of Purdue University*, Series No. 128, March 1967.

Research and Special Programs Administration, "Guidelines for Applying Criteria to Designate Routes for Transporting Hazardous Materials," Report No. DOT/RSPA/OHMT-89-02, June 1989 (revised version of Reference 1).

Smith, R. N., "Predictive Parameters for Accident Rates," California Division of Highways, 1973.

Urbanek, G. L., and E. J. Barber, "Development of Criteria to Designate Routes for Transporting Hazardous Materials," Report No. FHWA-RD-80-105, Federal Highway Administration, September 1980.

Evaluation of Hazardous Material Transportation by Rail

William H. Oderwald (1) and Mary Anne Sontag (2)

1.0 INTRODUCTION

ALK Associates Inc. of Princeton, NJ has developed a computer model, the Rail Hazardous Materials Routing System, which provides historical accident rate and population exposure statistics for rail routes under study for the purpose of transporting hazardous materials. This system is used by shippers, receivers and manufacturers of hazardous materials to evaluate current and proposed rail routings. In addition, the system suggests alternative routes which will minimize routing criteria such as accident rate, hazardous accident rate, route serviceability and population exposure.

The four major components of the Rail Hazardous Materials Routing system are the Princeton Transportation Network Model (PTNM) a computer based transportation network and geographic information and mapping system, the yearly Federal Rail Administration Accident/Incident Files, aggregated traffic statistics from the Interstate Commerce Commission Carload Waybill Sample, and census information from the Department of Commerce. This paper will describe the application of the three data bases to the PTNM along with an explanation of the function and philosophy of the PTNM. This method of analyzing hazardous material movements by rail can be extended to other transportation modes, and is discussed in this paper.

2.0 SOURCE AND RELEVANCE OF THE DATA BASE

The data base for this analysis has four components. They are the Princeton Transportation Network Model (PTNM), the Federal Rail Administration (FRA) Accident/Incident file, the Interstate Commerce Commission (ICC) Waybill Sample, and the 1980 Census population data. Each component is detailed in this section.

---

1. Vice President, ALK Associates, Inc., 1000 Herrontown Road, Princeton, NJ 08540.
2. Product Manager, ALK Associates, Inc., 1000 Herrontown Road, Princeton, NJ 08540.

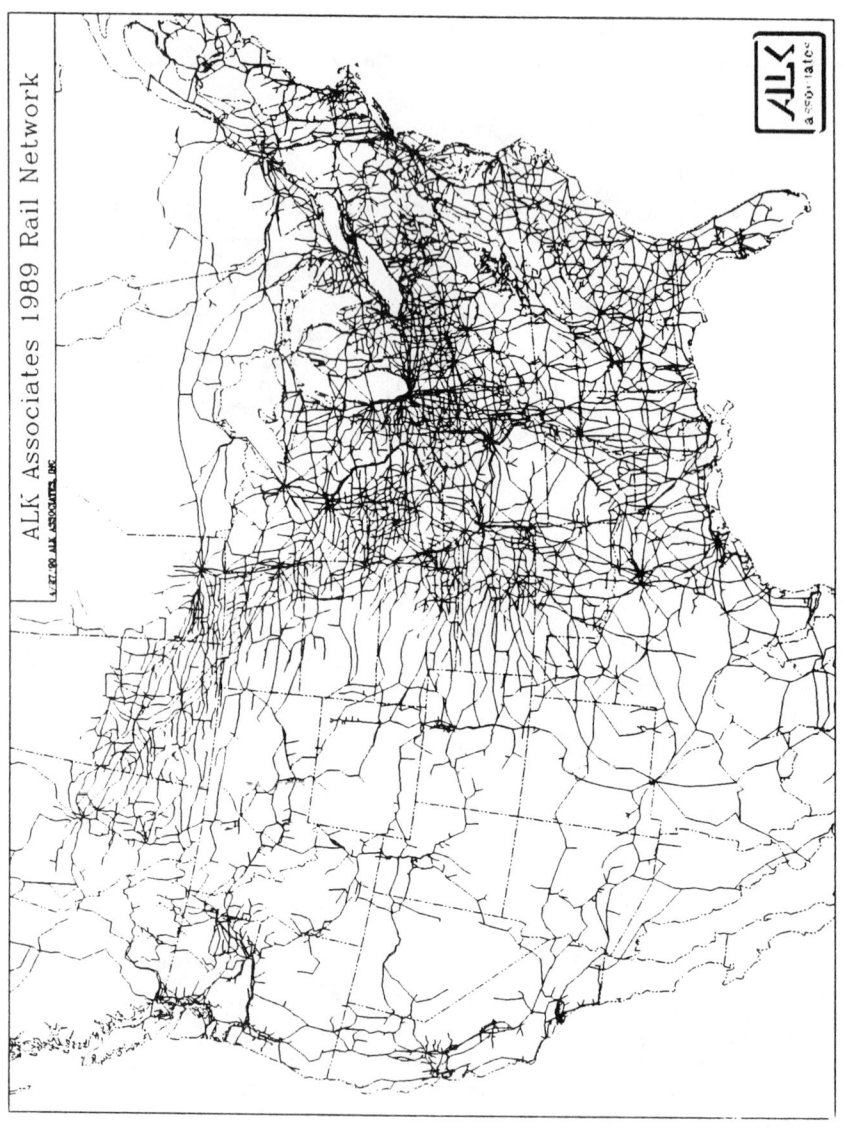

Figure 1

RAIL TRANSPORTATION EVALUATION 193

2.1 PRINCETON TRANSPORTATION NETWORK MODEL (PTNM)

   The rail network included in the Princeton
Transportation Network Model currently comprises 21,612
rail stations (nodes), representing 48,550 active freight
stations connected by 24,514 rail segments (links). The
network represents the freight railroads in the United
States, Canada and Mexico. A graphical representation of
the most recent version of the PTNM rail network in North
America is shown in Figure One. The original basis of
the network is the 1975 Federal Railway Administration
'503'. The network has been modified to include
additional detail in many urban areas, Canadian and
Mexican railroads, and Class III line-haul carriers.
Each carrier in the network has been adjusted over time
to reflect abandonments, mergers, spin-offs and new rail
lines. Figure Two shows the changes in the Illinois
Central Gulf Railroad Company (ICG) of 1981 to the
present day Illinois Central Railroad Company (IC) from
approximately 8,000 miles of track in 1981 to the IC with
about 3,000 miles in 1989.

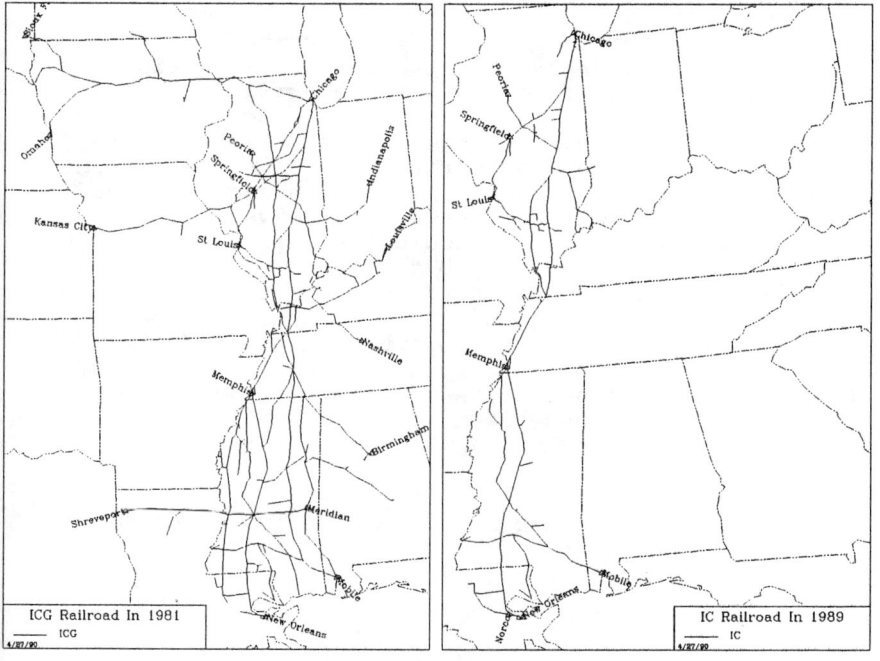

Figure 2

Node characteristics include geographic coordinates, tied to latitude and longitude, and the Standard Point Location Codes (SPLC) at each node. Links in the network are defined by their node endpoints, and have as attributes the owning railroad and those having trackage rights, the link distance, an indicator of track quality, and population density within a specified distance around the link.

Routing is accomplished by an algorithm which minimizes the sum of given weighting factors for the links along a route. Weighting factors utilized by this model include track classification codes, population density and whatever accident information has been ascribed to the links. Thus routes on the model may be generated which minimize route distance, accident frequency per traffic unit or population exposure.

2.2 FRA ACCIDENT/INCIDENT FILE

The Federal Rail Administration (FRA) Accident/Incident File is compiled annually by the FRA, and reports freight and passenger train accidents according to the rules found in the Accident/Incident Bulletin published by the Federal Rail Administration, Office of Safety. Currently, ALK is using accident statistics for the years 1986-1988, and has available statistics from 1975-1988. Although the reporting compliance for this file has come under fire recently, this FRA file remains the single best source of rail accident information. Deficiencies in this file are the relative inaccuracy or obscurity of accident locations (e.g. 'Columbus OH', 'Cabin 2') and the lack of specific commodity information for the load involved. As Figure Three shows, the number of accidents reported to the FRA declined rapidly from 1978, but has increased slightly recently, perhaps due to increased reporting compliance. Note that accident frequencies are location (node) specific. Figure Four is a graphic display of the freight accident locations reported to the FRA in 1988. The translation of accidents from node to link characteristics is described below.

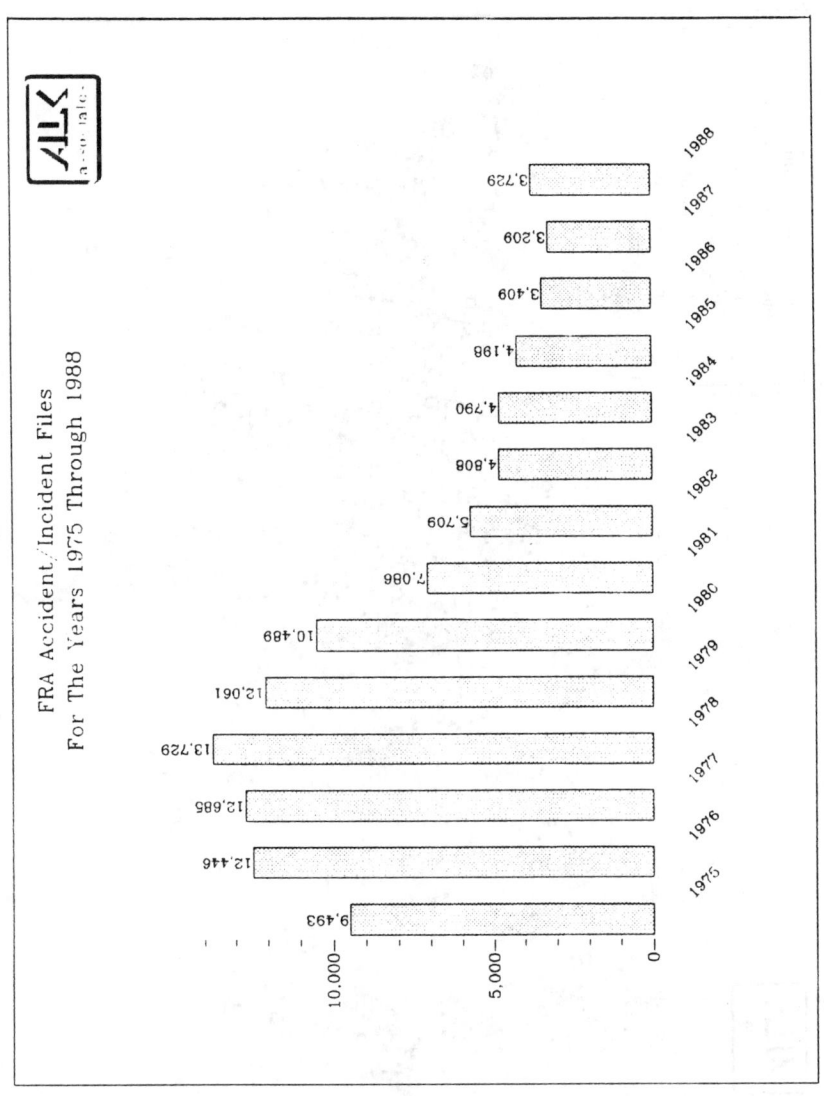

Figure 3

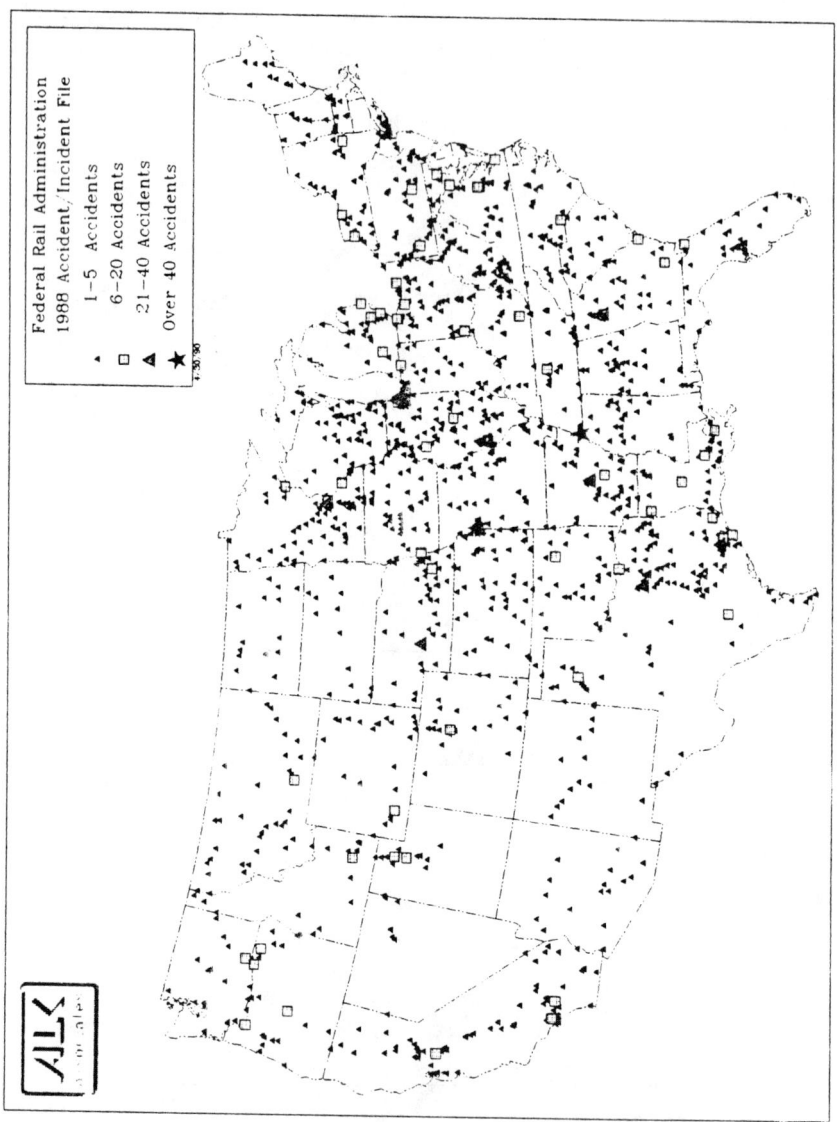

Figure 4

2.3 Rail Volume Data

In order to calculate accident frequencies, some measure of traffic volume is necessary. ALK currently uses an aggregate three year average of information ultimately compiled from the Interstate Commerce Commission Carload Waybill Sample (CWS) under the ICC's rules for aggregation of these data. The restrictions on the use of the CWS dovetail with the deficiencies in the data, that is, that origin and termination information is often aggregated, and particular routes travelled (yard to yard) may change daily. For detailed local studies, this volume information must be supplemented by information on particular traffic patterns and yard procedures. The origin-destination-volume information is assigned to the PTNM to derive representative link volume characteristics.

2.4 POPULATION DATA

The population data originally used in the Rail Hazardous Materials Routing Model was developed by the Transportation Systems Center in conjunction with ALK and was based on the 1970 Census. ALK used the 1980 Census to factor these data to reflect growth (or contraction) in population from 1970 to 1980. Upon the release of the 1990 Census data ALK will assign the Block Group and Enumeration District data to the current PTNM rail network through the use of interlocking longitude and latitude characteristics. For most present studies ALK considers the population within a one-mile band around a rail link. For some purposes, however, such as the transportation of nuclear weapons or wastes, wider bands have been used. In all cases, it is important to remember that U.S. Census data are largely residential, and that time-of-day must be considered in any estimation of population exposure.

Figure Five depicts ALK's 1989 rail network with a three mile band of population density surrounding each link in the network. With more current data and a newly developed algorithm ALK will be able to generate population density for any size area surrounding a set of links.

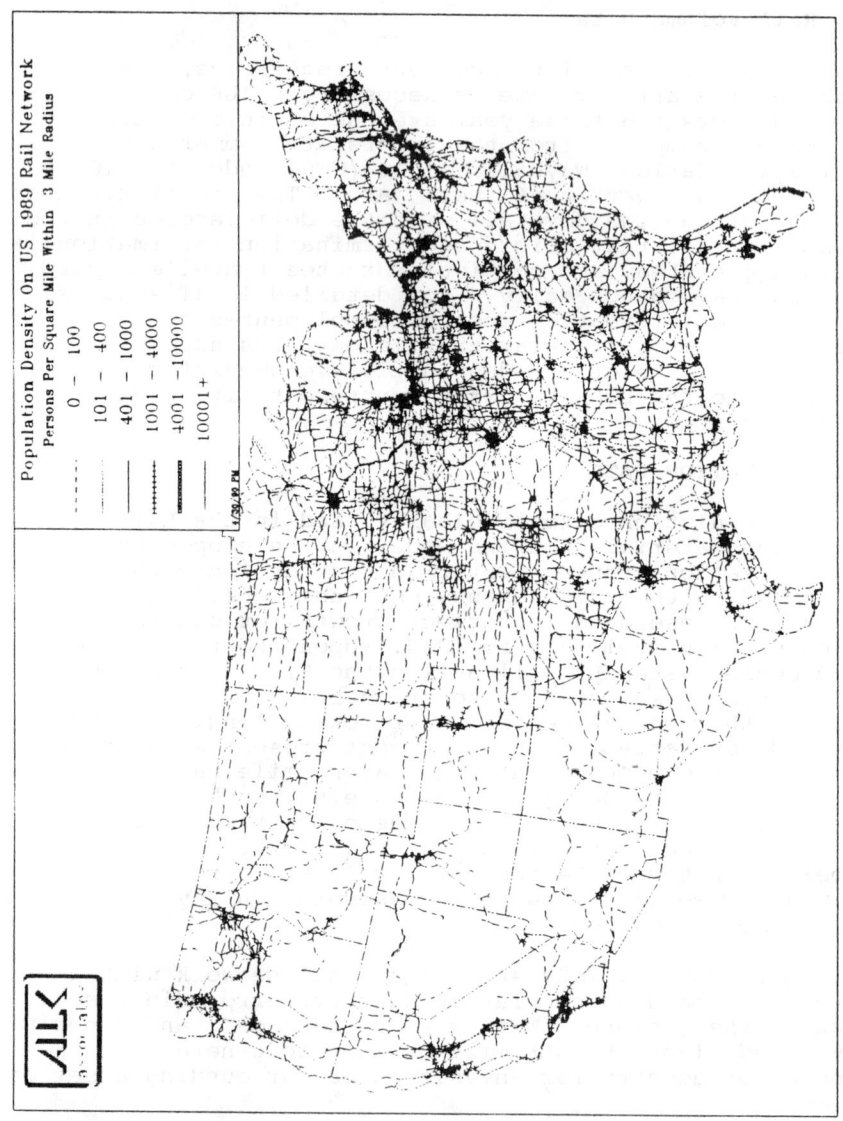

Figure 5

3.0 DATA BASE UTILIZATION

This section discusses the three steps followed in preparing the input data for use in the Hazardous Material Routing Analysis.

3.1 GEOCODING ACCIDENT FILE

The first step in the Hazardous Material Routing Analysis process is the translation of city and state locations in the FRA Accident/Incident file to the node numbers of the PTNM. For this step translation tables are used which match Federal Information Processing Standard (FIPS) codes to state abbreviations, and match city and state names to PTNM node numbers. In an initial run the FIPS codes are translated to state abbreviations. The city and state locations that did not translate to PTNM nodes are then checked by hand. Typically, untranslated locations are the result of misspellings. Figure Six shows accident locations in a twenty-five mile radius around Corinth MS. This display technique, developed in conjunction with the U.S. Air Force, is useful for the determination of accident and/or population 'hot spots' within a give radius around an area.

3.2 ASSIGNING ACCIDENTS TO LINKS

Once the accident file has been geocoded, the next step in the process is the assignment of the accidents to PTNM links. Since accidents are characteristics of a node, and routing is done on the basis of links, the accidents must be assigned from nodes to the links which are defined by the nodes. Two important characteristics of a link is the amount of traffic that traverses that link or the link volume, and its nodal endpoints or connecting points. Since, in the final analysis, the model generates an accident per volume ratio, accident locations need to be assigned as link characteristics. The algorithm used to do this looks at each individual accident location and assigns 0.5 accidents to every link going into or out of that location for every accident that took place at that location. For example, if there were four accidents in Anson, TX in a given year, every link with Anson as an endpoint would have 2 accidents assigned to it. In addition if Abilene, TX had 1 accident, every link with Abilene as an endpoint would have 0.5 accidents added to it. Since Anson and Abilene are endpoints of one link, this link would then have 2.5 accidents assigned to it (2 from Anson) + 0.5 (from Abilene). The sum of the accidents on any two links going into or out of any node equals the number of accidents at that node. Therefore, any routes going through that node will be assigned half of the number of

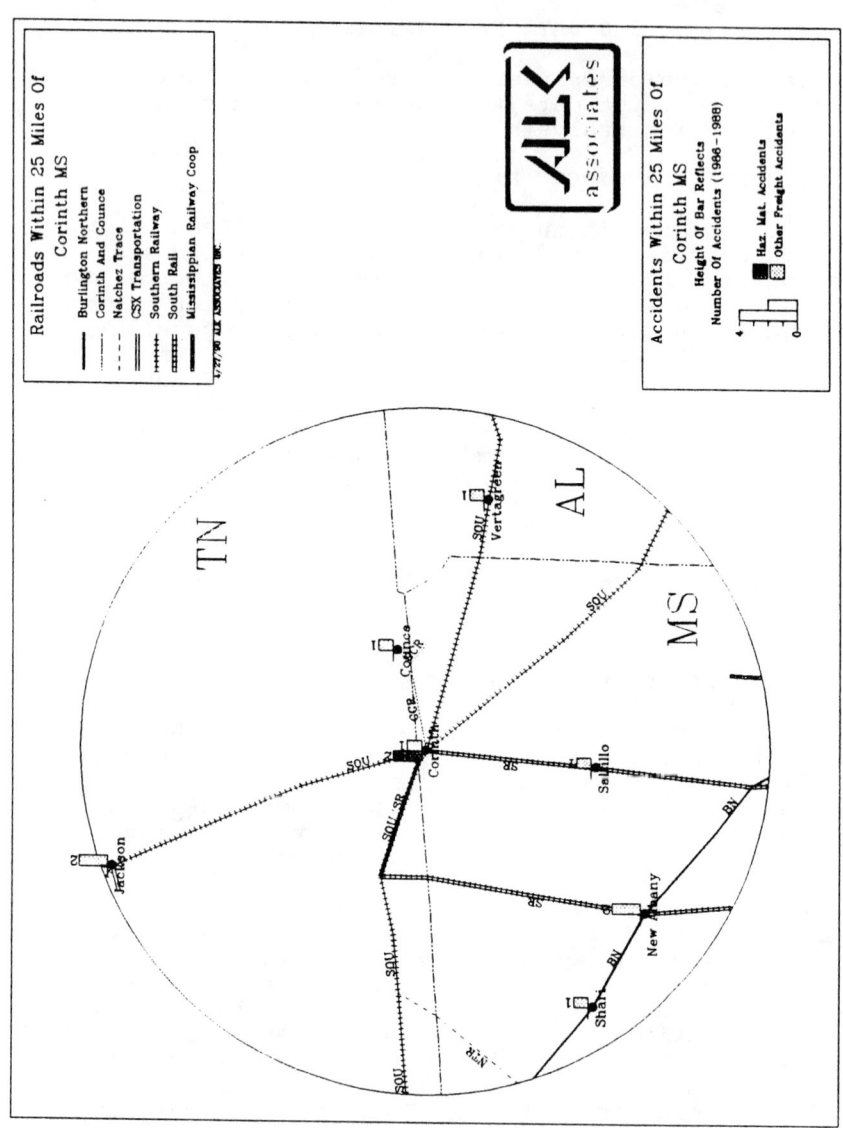

Figure 6

accidents going in and half of the number of accidents going out of the node. See Figure Seven for an illustration of this concept.

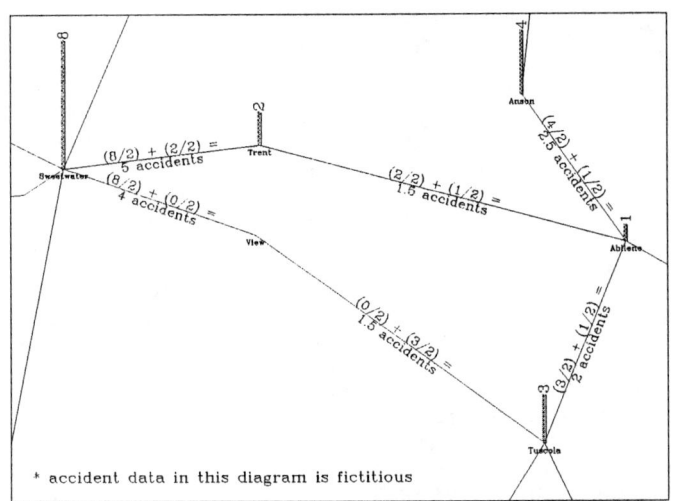

* accident data in this diagram is fictitious

Figure 7

3.3 PROCESSING ROUTES

The Hazardous Materials Routing Package uses two routing models developed by ALK. The Intracarrier model determines the likely service route between any two points on a single railroad. The Quantanet Intercarrier model is used to generate the benchmark routes (i.e. minimum impedance, minimum population, minimum accidents, minimum hazardous material accidents). It considers all originating railroad/terminating railroad combinations between an origin and terminations under study.

The benchmark routes for minimum population exposure and minimum accident exposure are often impractical routes that would require special train service. However, the characteristics of these routes provide a useful base against which to compare the characteristics of the practical routes.

4.0 APPLICATIONS OF THE DATA BASE

Once the data base is assembled from the four components (the FRA Accident/Incident file, the volume statistics, the PTNM and the population data), the next step is to process the routes. The Hazardous Material

Routing Analysis provides origin to destination information for each of the routes listed below:

* Base Route
* Minimum Impedance Route
* Minimum Population Route
* Minimum Accident Route
* Minimum Hazardous Accident Route

The Routing Analysis returns accident information, traffic volumes, and population data for the links along each route. The first of these five routes is the Base Route. This route is usually supplied by the client as the route currently used for this particular origin-to-destination pair. The Routing Analysis returns accident information, traffic volumes, and population data for the links along each route.

To generate the remaninig four routes the Quantanet Intracarrier Model is used. This model looks at all possible routes from Point A to Point B, considers all possible rail carrier combinations and chooses each route based on each of the four criteria. The criteria used for the Minimum Impedance (Min Imp) Route is dependent on track quality, distance, and the unlikeliness of carriers to interchange unnecessarily. The Minimum Population (Min Pop) Route is determined by the population along each link; therefore track quality, distance and interchanges between carriers are not considered. The Minimum Accident (Min Acc) Route is selected by determining the route with the smallest accident per volume rating. The last route generated is the Minimum Hazardous Accident Route (Min Haz). This route is selected on the smallest ratio of hazardous material accidents per hazardous material volume on each link.

After the five routes have been identified the algorithm generates accident summary reports, statistical tables showing population impacted, mileage, and accident frequency for each route. Another output of the Hazardous Material Routing Analysis is route maps. There are six maps produced for each origin-destination pair. The routes generated by ALK's Hazardous Material Analysis can be used as a routing guide and statistical input for risk assessment decisions. However, if someone is particularly concerned about a shipment that person should confer directly with the railroad in question.

5.0 EXTENSIONS OF CURRENT RAIL METHOD

Since rail is not the single mode of transportation for hazardous material, ALK also has alternate routing analyses for highway transportation and looks to the future for waterways, pipeline, and airline analyses.

Currently an extensive highway network including over 54,000 links (or road segments) and over 150,000 place names is included as part of the PTNM. This network includes over 500,00 miles of major and secondary roads in North America.

Population figures are also applied to the U.S. highway network in the same way they are applied to the rail network. Currently we are using the the figures from the 1980 Census at the block level. As with rail, the 1990 Census figures will be applied as soon as they are available.

Accident figures are taken from the Bureau of Motor Carrier Truck Accident File (TAF) and individual state vehicular county accident counts. ALK has the TAF for the 1981 through 1988, however for the years 1981 through 1984 there are many invalid fields. We have found a significant improvement in the files starting in 1985. Although we feel the data is acceptable for the last four years, in our analysis we use the current three years (1986-1988).

Traffic volume statistics were the hardest to find and implement. Since there is no uniform and reliable traffic volume base on a national level, we have contacted the 48 contiguous states and received some form of average daily traffic volumes. The volume statistics came from several different years and in various formats ranging from books of road segments with directional vehicular and truck data to colored flow maps in which traffic volumes were broken into as few as five categories and without truck data.

Over several months, ALK entered this data and normalized it through a factoring scheme to an average 1987-1988 data format. Although the reliability of this data is somewhat suspect, ALK feels it is the best data available at this time.

With this data base, ALK can generate minimum impedance routes, minimum distance routes, minimum population routes, an historical route, and a minimum accident route. Each route processed is detailed with the name of the road segment, distance travelled on that segment and the population impacted should an accident occur along that road segment. One of the major problems with routing on the highways is the lack of a central clearing for average daily traffic (ADT) counts. Some of the criteria for routing hazardous materials on the highway that ALK is considering are emergency response routes, time of day routings, and avoidance of certain areas like water supplies.

Another problem of highway routings is the complexity of our nations roadways. Right now many state legislators are specifying which roadways in their state can be used for hazardous material shipments, resulting in a disconnected roadway system. Modifications to this nation wide system are continually being made. However, when the states and federal government decide on a final continuous road system, ALK expects to maintain this network in addition to our general highway network.

BIBLIOGRAPHY

1. Accident/Incident Bulletin, No. 149, Calender Year 1980. Washington, DC: Department of Transportation, Federal Railroad Administration,Office of Safety, Appendix A, June 1981.

2. Census of Population and Housing, 1980: Master Area Reference File (MARF) 2 (machine-readable data file) / prepared by the Bureau of the Census, Washington: The Bureau (producer and distributor), 1983.

3. Census of Population and Housing, 1980: Master Area Reference File (MARF) 2 Technical Documentation / prepared by the Data User Services Division, Bureau of the Census, Washington: The Bureau, 1983.

4. Kornhauser, Alain L. "Comparative Analysis of the Flow of United States Railroad Freight Traffic for the years 1973, 1974, 1975." Report to the Interstate Commerce Commission, Princeton, NJ: Princeton University Transportation Program, 1977.

5. Oderwald, William H., and Sontag, Mary Anne "Hazardous Material Routing Analysis". Presented to National Association of Regulatory Utility Commission, Princeton NJ: ALK Associates, Inc., 1988.

6. The Official Railway Guide. North American Freight Service Edition. New York NY: National Railway Publishing Company, Vol. 117, July-August 1986.

7. The Official Railway Equipment Register. New York NY: National Railway Publishing Company, Vol 98, Oct 12, 1983 et. seq.

A Risk and Vulnerability Assessment Approach for
Selecting Routes: Case Study of Hazardous
Waste Transportation in Arizona

K. David Pijawka, A. Essam Radwan, and J. Andy Soesilo[1]

Abstract

The study objective was to provide an approach to selecting routes to a proposed hazardous waste treatment and storage facility, based on a risk and vulnerability assessment. Probability estimates were based on national transportation accident rates for hazardous material carriers, local accident characteristics, and the magnitude of projected shipments to the facility. Release incident probabilities resulting from equipment failure in nonaccident situations were also considered. Population risk values were determined in order to compare each route segment. The vulnerability component evaluated the transportation risks against the capabilities of the area to respond to and mitigate the hazards.

Introduction

This study was conducted to provide information on the risks of transporting hazardous waste (HW) to the proposed Arizona Hazardous Waste Management Facility to be located near Mobile, Arizona. A study area was defined in which route segments were identified for risk analysis (see Figure 1). This study was motivated in part by the provisions of A.R.S. 36-2806 requiring the Arizona Department of Health Services (ADHS) to issue regulations to specify east and/or west access to the

---

[1]K. David Pijawka, Assistant Director, Center for Environmental Studies and Associate Professor, School of Public Affairs, Arizona State University, Tempe, AZ 85287-1201; A. Essam Radwan, Associate Professor, Center for Advanced Research in Transportation, Arizona State University, Tempe, AZ 85287-6306; and J. Andy Soesilo, Adjunct Professor, Arizona State University, Tempe, AZ 85287 and Planner, Hazardous Waste Section, Arizona Department of Environmental Quality, 2655 E. Magnolia, Suite 2, Phoenix, AZ 85034.

facility. This assessment addressed the risks of transportation accidents to the site.

The risk assessment consisted of the following analytical elements:

1. The development of a probabilistic accident release equation to measure the level of risk -- the probability of occurrence of an accident event resulting in release of HW and the population that could be at risk from such an event for every route segment in the study area.

2. The collection of data of route segments included length, average daily traffic volumes, prevailing accident rates, and population residing 1.5 mi on each side of the routes.

3. The application of accident release probabilities to a "capacity shipments" scenario. That is, the evaluation of accident probabilities was based both on the facility operating at full capacity and the utilization of <u>any</u> route to the facility. The risk analysis would not be based on predetermined routes,

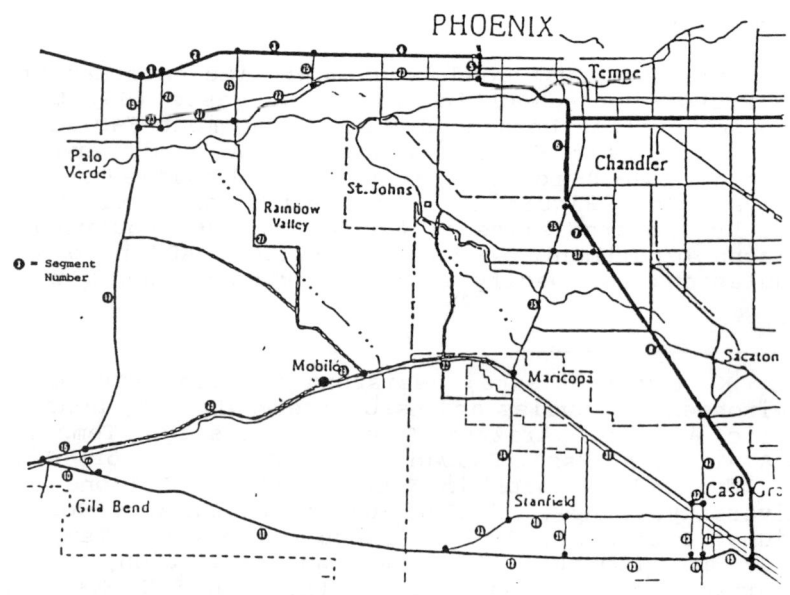

Figure 1. System of Route Segments

because any and all routes could be utilized for shipments to the facility with the number of shipments being the same for all routes. The analysis can be viewed as a "worst case" scenario.

4. The probability risk estimations above were based on release following accidents. However, release of some HW contents may occur because of equipment failures that are unrelated to accidents. The probability of these releases was established as well as for the routes.

5. The assessment of the severity/consequences of potential releases was undertaken. This included comparing routes by population risk values. In addition, potential accident "severity" was evaluated on the basis of the degree to which containers would release the HW contents following accident events.

6. A "hazard vulnerability" evaluation was completed. The capabilities of emergency response organizations and preparedness resources were measured against the risk. Hazard reduction efforts were also recommended in two broad areas -- mitigation and preparedness enhancement activities. As part of the vulnerability assessment, sensitive public use facilities were identified.

The study was not conducted to specifically recommend routes to the facility. Rather, information was provided on accident release risks and population on a route-by-route basis that will assist in making routing decisions.

Approach to the Risk Analysis

Identification of Route Segments. The study area is delineated by Interstate 10 on the north and east, Interstate 8 on the south, and State Highway 85 on the west. In total, 42 segments were identified in the study area, as shown in Figure 1. It is important to note that some of these routes were not "all weather" routes, were not paved, and were not up to a standard that was compatible for HW shipments. However, for the risk analysis, the study assumed that these routes would be upgraded to meet the standards. Therefore, risk estimates were developed for routes that would not presently meet the regulatory standards for HW transportation.

Population data was collected from the 1980 census tracts for distances 0.5 mi, 1.0 mi, and 1.5 mi on either side of the route segments. The 1.5-mi population figures were used in the risk analysis as being the most conservative. The use of the 1.5 mi was based on U.S. Department of Transportation's evacuation distances for certain chemical spills. Since we can expect some changes in population since 1980, the population-at-risk was considered underestimated to a small degree. Major changes in population since 1980 along the rural segments were observed in a "windshield survey" that was also undertaken.

Number of Shipments. The first step in the risk analysis was to determine the daily number of trucks that would be transported to the site when the facility will be operating at full capacity. Communication with personnel from the waste management firm who will operate the facility indicated that the capacity of the facility will be at 78,000 tons/yr over a 10-yr period. Three types of trucks are expected to serve the facility: (1) bulk tankers; (2) bulk solids trucks (covered dump trucks); and (3) trucks carrying drums.

For truck types 1 and 2 it is expected that the maximum carrying capacity ranges between 4,000 and 5,000 gal. This results in a maximum truck capacity of 40,000 lb. Therefore, the expected capacity is 20 tons per truck when full. As for type 3 trucks, it is expected that 70 drums would be carried per trailer. Assuming a drum capacity of 500 or more pounds, the resulting truck capacity would be about 20 tons. Thus, at the shipment rate of 78,000 tons/yr, an average of 14 trucks/day would deliver HW to the facility, regardless of the truck type. However, in addition to the incoming HW shipments, recovered products (solvents and calcium chloride) would also be sent out from the facility. An estimated 13,500 tons/yr are expected to be shipped from the treatment facility. Another 4,200 tons/yr of chemical reagents can also be expected to enter the facility to be used in treatment processes. In sum, an average of 19 shipments can be expected daily based on a 250 working-day year. The "capacity" scenario, therefore, assumes that each segment will have 19 trucks daily.

Risk of Accident Release Incidents by Route Segment

The objective of this section of the analysis was to derive the probability of a release of HW resulting from an accident along each of the route segments. It was based on an approach using both national data on HW accident rates and accident rate data for the local route segments. The expected number of accidents resulting in release of HW along individual segments is given by the following equation:

$$FA_i = IA(AR_i/AAR)(V)$$

where: FA_i = annual frequency of release incidents due to HW truck accidents on individual route segments; IA = national rate of release incidents due to truck accidents/HW truck-mi, nationally; AR_i = annual rate of vehicular accidents/vehicle-mi on individual route segments; AAR = average annual rate of vehicular accidents for the study area; and V = annual volume (truck-mi) of HW traffic on individual route segments going to the proposed HW treatment facility.

Although this is but one of many possible equations which could have been constructed for computing a probability estimate of FA_i, this has the advantage of relying upon readily available data. Specifically, HW accident and incident release data on the number of HW shipments to the facility and accident data by individual segment. This is a conservative risk approach because the national rate often would be higher than the local Arizona rate due to inclement weather conditions elsewhere and the larger density of HW traffic in California and the Northeast. In addition, the maximum number of trucks going to the facility is assumed. Basically, the analysis begins with a national release rate (IA) which is adjusted to local accident rate conditions by assigning local rate weights to the segments over which HW shipments could take place. This adjusted rate is then multiplied by the maximum annual volume of HW shipments (vehicle-mi) on each segment being considered. This yields the projected annual frequency of release incidents.

We have the following estimates for the above components:

1. $$IA = \frac{8.5 \times 10^{-6} \text{ release incidents per shipment}}{50 \text{ mi average shipment distance}} =$$

   1.7×10^{-7} release incidences/HW truck-mi

   This amount is derived from Department of Energy reports on HW tank trucks in the U.S. (Rhoads et al. 1978) and is supported by other studies on HW tank trucks (F. G. Bercha and Associates 1980).

2. AR_i/AAR. The value for this ratio is based on the prevailing accident rate on individual segments to the average accident rate for the study area as a whole. Accident rates are defined as the number of vehicular accidents/million vehicle-mi annually. This value permits the differentiation of route

segments and provides consideration of local conditions.

3. V = 19 HW trucks/day(250 working days/yr)(individual segment lengths in mi)

The basis for the volume estimate of the shipments was based on the 19 trucks/day estimation noted above. The risk analysis is based on 4,750 trucks/yr carrying HW on each of the 42 identified segments within the study area. A risk analysis based on the above assumptions was conducted for each route segment.

The largest probability of HW release occurs on the "outer" segment (on link 23). The release probability is estimated to be 18.71×10^{-4}. The implication of this probability is that we may expect 1.8 accidents resulting in release every 1,000 yr based on 4,750 shipments annually. Therefore, the largest probability of an accident release will likely be very small. However, while the probability shows that such an event will occur once in every 3.68 million shipments, this event can occur at any time. Probability refers to the chance of occurrence of an event/unit time and does not refer to a prediction of occurrence. Therefore, preparedness measures are important to develop to meet this low probability event occurrence.

Our principal concern was with those route segments that were located within what is designated as the "inner segments," identified as segment numbers 27 to 42. These are generally rural routes. The data show that the probability of release due to an accident ranges from 1.83×10^{-4} (segment 39, south of Stanfield) to 9.07×10^{-4} (segment 31, Maricopa to Casa Grande). This represents one accident resulting in a release for every 7.1 million HW shipments. Maricopa Road, the major paved access road to the facility, represents risk probabilities of release slightly above the average for the inner segments. Segments that exit I-8 to Stanfield (segments 33, 38, 39) and Casa Grande (segments 40, 41) represent routes with slightly less than the average risk of release events. The route segment from Casa Grande to I-10 (segment 42) also shows a higher than average risk rating. Except for low probability segments (33, 38-42) the variation in the probability of release is quite small. To reiterate, each segment represented a worst-case scenario and the risk analysis found that release probabilities were small.

Figure 2 shows the distribution of accident release probabilities by route segment. In general, the inner segments represent the lowest to mid-level probabilities of release while some of the outer segments represent the largest probabilities. The lowest probability category was found in segments 33, 38, and 39 (leading to Stanfield from I-8) and two routes south of Casa Grande.

Route segment 17 (State Highway 85) has an estimated accident release probability of 13.39×10^{-4} and segment 6 (I-10 between central Phoenix and the Maricopa Road Exit) has an accident probability of 12.91×10^{-4}. These two segments reflect accident probabilities that are larger than any of the rural inner segment routes. Segment 11 (I-8 between Gila Bend and the Stanfield Exit) was estimated to have an accident release probability of 8.68×10^{-4} and falls in the mid-probability category which is higher than the average probabilities of the inner segment routes.

Population/mi determinations for each inner segment were made. The largest population/mi concentrations were found on segments 32, 40, and 41 (routes leading to and through Casa Grande, with over 3,000 persons/mi). With a concentration of 554 persons/mi, segment 42 (north of Casa Grande to I-10) has the next largest population. Outside of the Casa Grande area, population/mi along route segments ranges from zero persons (Riggs Road) to 61 persons/mi. The average population-risk value for the study area was found to be 0.39. Routes through and around Casa Grande averaged 1.88 on the population risk value.

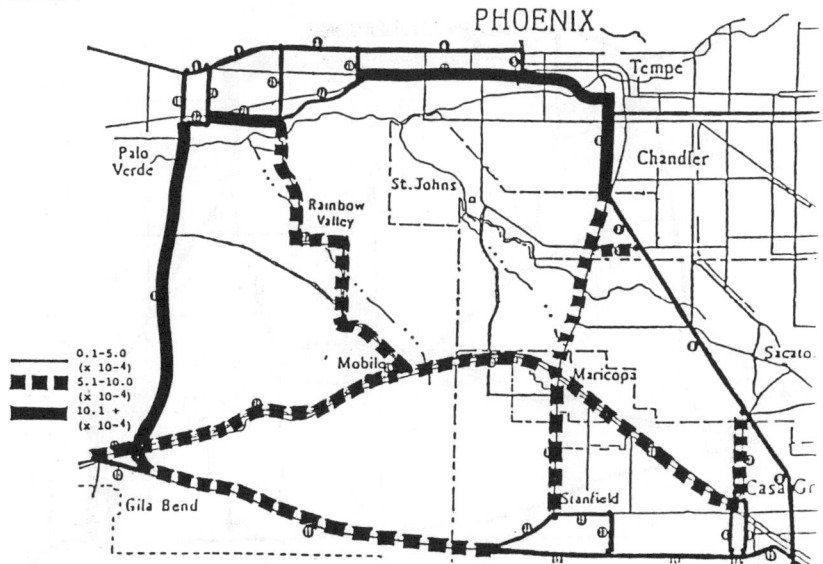

Figure 2. Probability of Accident Release Incidents by Route Segment

As shown in Figure 3, the population risk values are categorized into three classes. The area around Casa Grande, for example, represents the largest population risk factor. This rating was largely determined by the high relative population density along those segments despite a relatively low accident release probability. Except the Casa Grande area, all other inner segment routes are in the lowest population risk category. All segments within the lowest population risk category present extremely low risk, both in terms of event release probabilities and population at risk.

Release Incidents Due to Equipment Failures. This study also investigated the release incidents that may be due to truck equipment failures. When such incidents occur they result most often in very minor releases, which can be suppressed and cleaned-up much more readily than accident-related spills. Seldom have these resulted in human fatalities. However, equipment failure releases can be expected to occur more often than releases resulting from accidents. Equipment failure releases are not related to local road conditions but are related to vehicular age, engineering standards, container types, failure modes, and causes codes used in the Materials Transportation Board incident reports. These releases can be exemplified by a small amount of material leakage due to a valve malfunction while the vehicle is in motion.

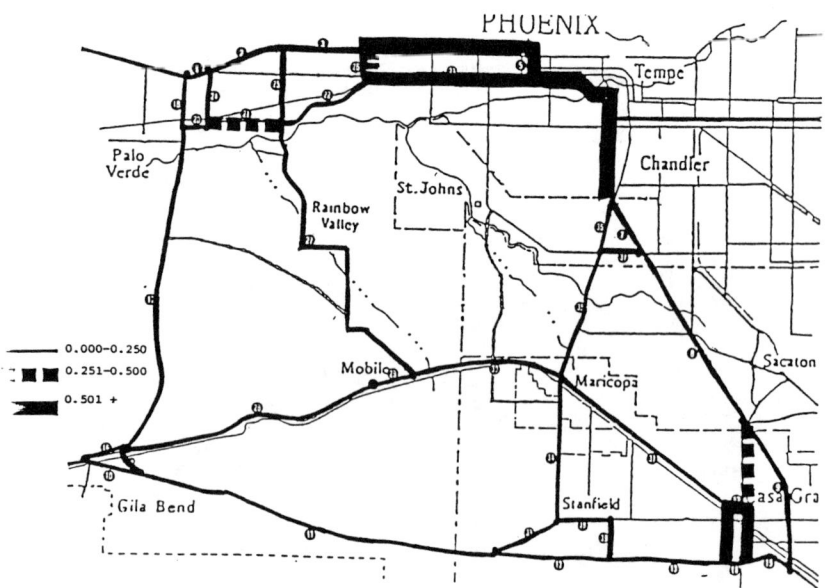

Figure 3. Population Risk/Mile by Route Segment

Release rates from equipment failures are not directly related to accident rates. Variation in these occurrences are related to the number of vehicle-mi of HW shipments. That is, the longer the distance traveled by a HW truck on an individual route, the greater the likelihood of such incidents. However, good condition and maintenance levels can reduce these small release incidents. Inspection programs and more stringent standards are controllable measures that can result in lower release rates.

Based on an analysis of release incident rates, release rates of 2.2×10^{-7} and 2.4×10^{-5} failures/truck-mi are given for tanks and drums, respectively. Bulk cargos in dump trucks, which are the only other relevant form of shipment have not been analyzed (IDF, Inc. 1984).

The equation for the annual frequency of release incidents is as follows:

$$FE = [IE, T(V_T)] + [IE, D(V_D)]$$

where, FE = annual frequency of release incidents due to equipment failures on the local access route; IE,T = total failure rate/truck-mi for tanks; IE,D = total failure rate/truck-mi for drums; V_T = annual volume (truck-mi) of HW truck traffic in tanks on the local routes to the facility; and V_D = annual volume (truck-mi) of HW truck traffic in drums on the local segments to the facility.

Based on the assumed shipment rate of 4,750 trucks/yr and the national release rates of equipment failure, the analysis found that on a mile basis the probability of an equipment failure release was 1 in 17.4 yr for any route segment. This release rate is length dependent; the longer the route the larger the probability estimation. If the HW industry moves toward greater tank containerization (and there is some evidence to support this trend) the probability of releases due to equipment failures will consequently decrease accordingly.

Vulnerability Analyses for the Transport of Hazardous Wastes to the Mobile Site

From an emergency management perspective, the vulnerability of a community in a region is determined by the extent to which it is exposed to a hazard and its ability to respond or mitigate the effects of that hazard on both the physical and social environment. This part of our analysis focuses on the current resources available to respond to an accidental spill of HW while in transit to the Mobile site.

The term "resources" refers to both the availability of personnel and equipment to manage such an incident, to neutralize its effects on the physical environment (including groundwater, surface water, and soils), and to protect humans from potentially harmful effects on individual health as well as collective well-being. Personnel resources include fire fighters, police and sheriff's department officers, those trained in emergency medical procedures, and others with emergency management responsibilities at the city and county levels of government. Equipment includes those physical resources that these personnel have at their disposal to adequately respond to HW spills within their jurisdictions without imperilling their own safety.

The analysis will also consider an assessment of the "systemic" resources available to the involved jurisdictions along the potential routes into the Mobile site. System resources include the extent to which emergency response personnel have been trained to respond to such incidents, to the interorganizational and intergovernmental planning that has been undertaken to respond to emergency situations, to the existence of interagency emergency communication systems, and to the medical infrastructural resources that exist to protect human health in the event of exposure to HW.

The adequacy of these emergency response resources will be evaluated in relationship to the population "at risk" from exposure to possible spills or atmospheric releases. For each route segment, we have identified residential populations at risk within 1.5 mi, on either side of the transit corridor. In addition, sensitive areas have been identified within each of these zones. A sensitive area refers to public use structures (e.g., schools, auditoriums, sports facilities), emergency response facilities (e.g., fire stations, hospitals, sheriff's offices), and commercial centers (e.g., shopping areas and large employment centers). These areas represent aggregations of the population that could become temporarily endangered due to their density, thus creating greater need for the availability of response and mitigation efforts.

For the first part of the vulnerability assessment, interviews were conducted with emergency response organizations in 14 communities. Of the organizations contacted, 15 fire departments, 5 police departments, and 5 sheriff's offices responded to our request for information. Fire chiefs, police chiefs, or station commanders (or their identified designees) were used as knowledgeable informants about the departmental resources available to respond to a HW transportation incident.

For purposes of this analysis, it was assumed that more than just the "first response level" training and adequate equipment availability for HW incidents would be required to prepare fire fighters and other emergency response organizations to deal with situations involving the accidental release of HW. The state's role is to provide HW planning coordination and it has established state hazardous materials response teams. However, local response capabilities are the key to effective response.

Of the overall study communities, only three are staffed totally by paid personnel (of these, only Casa Grande is within the internal study area), several totally by volunteers (four of which are within the internal study area), and five by a mixture of both paid and volunteer fire fighters. Only 12% of the fire fighters in the overall study area are available along the internal route segments. Of these, most are either restricted to the city limits of Casa Grande, the Gila River Indian Community (GRIC), or not quickly available to the Maricopa area because of a lack of paved roads from the north and west.

With the 14 communities in the study area, there area a total of 1,570 fire fighters. Of this total, 85% have had at least "first responder" HW training. Along the internal route segments, however, only 15% have been exposed to this training. While this represents an improvement from a few years back when the percentage of fire fighters exposed to some type of HW training was lower, the extensiveness of that training varies greatly. Most fire fighters outside of the Phoenix metropolitan area, especially volunteers, have only had introductory courses that emphasize safety aspects for early responders to the incident rather than training in how to neutralize the hazard.

The distribution of equipment resources along the possible route segments is also unevenly distributed to handle HW incidents. Along internal route segments, no fire department has special HW equipment available to handle incidents, with the exception of Casa Grande. However, the Casa Grande area (represented by route segment 32) has the largest number of residents at risk along any of the possible route segments, resulting in the highest population risk factor in the study area.

All departments which have some responsibility for route segments have some emergency medical personnel, primarily fire fighters with emergency medical training. Along internal route segments only Casa Grande has paramedics, however. While some rescue vehicles are available within the internal study area, the community likely to sustain the highest volume of HW transport vehicles regardless of route selected -- Maricopa -- has neither local ambulance service nor emergency hospital facilities.

Another source of HW response resources derives from police and sheriff's departments' personnel who serve the communities along these route segments. Attention should be focused on sheriff's stations located in communities along the route segments since they constitute first responder resources which might be involved in the assessment and management of initial response efforts. There is a striking scarcity of additional personnel resources that can be anticipated in the event of a HW incident, especially within the internal study area. Only two police departments are available on internal route segments (Casa Grande and GRIC), and their response capabilities are largely limited by their jurisdictional boundaries. Two *very* small sheriff's stations exist to cover internal route segments -- one in Maricopa and one in Stanfield -- and one only slightly larger office in Gila Bend. Since the geographic areas which the officers assigned to these stations are extensive, their availability to assist in response efforts would certainly be delayed and their capability to respond effectively would tax their existing personnel resources severely.

While this assessment determined that training in HW incidents has become more frequent for fire department personnel, there is still a need to expand the number of trained personnel and to enhance the quality of their training. This is particularly important for departments such as Maricopa, Sacaton, Stanfield, and Ak-Chin. Specialized HW equipment is not routinely available nor do the rural areas have sufficient resources for medical transportation in the event of HW accidents.

The vulnerability analysis compared the resources of emergency response organizations (particularly fire departments) with sensitive areas and areas with high population risk factors. By doing so, we find that three areas are particularly vulnerable in the event of a HW incident: (1) Casa Grande (segments 31, 32, 40-42). Although this city has the best HW resources within the internal study area, it also has the highest risk factor due to accident rates, population density, and commercial clusters along route segments. (2) Interstate 8 and lower Maricopa Road (segments 11-13, 33, 34). No fire department is responsible for responding to HW incidents in this area. Although the segments along the interstate have low population risk factors (due to low population density rather than to low accident rate), effects on the physical environment should be considered. In addition, segments 33 and 34 have relatively low population risk factors, but both have specific sensitive areas that should be given consideration for protection. This is especially important since no fire department is responsible for the area. (3) Mobile (segments 29, 30). Although population risk factors are very low, location of the elementary school on the route that all HW transport vehicles would take must be considered, especially since no emergency response agencies directly serve this area currently.

Hazard Reduction Activities

Hazard reduction activities are often conceptualized as consisting of two critical types of efforts -- mitigation and preparedness enhancement. Mitigation activities are those that will result in avoidance of adverse spill events, in the reduction of the frequency of accident occurrences, or in lessening the potential impacts of such events. Preparedness enhancement refers to those activities that increase the communities' capabilities to respond to a hazardous transportation event to protection public safety as well as the physical environment.

<u>Mitigation.</u> The following are recommendations to be considered for reducing the risk of accidents and their consequences in transporting HW to the proposed facility.

1. Risk Reduction. The risk and vulnerability analyses provide information on HW release probabilities and population risk factors to help in the designation of routes to the facility. Overall, the risks of accident release events are extremely low, but these can be further reduced by selecting a route system for these shipments. Such a route system should consider and incorporate the places of origin of HW shipments because the allocation of shipments along various routings rather than one designated route reduces the overall risk of transporting HW. Consideration should be given to the routes for HW shipments emanating from the northwest. For northwestern-originated shipments, Highway 85, I-8, segments 30, 33, and 34 should be considered as a route. This route would minimize population-risk and would avoid the high concentration of population in Phoenix.

2. Vehicular Inspections. The risk analysis presented above was based on accident release probabilities. While release probabilities are related to accidents and are strongly associated with road geometrics and traffic volumes, HW release rates are also the result of equipment failures. It is important that an effective state vehicular site inspection program be established for HW vehicles. An inspection program would tend to reduce these failures by assuring truck maintenance and safety standards.

3. Route Upgrading. An inventory of the major washes obtained from the U.S. Geological Survey topographical map, indicated that eight inner route segments are being crossed by these washes. These segments are 27-31, 35, 40, and 42. Future considerations should be given in routing HW shipments to the site during flooding periods.

   The upgrading of roads/infrastructure to meet the annual demands of HW shipments must be considered. Currently, segments 27-29 and part of 30 are not paved (dirt-top type) and cannot presently sustain truck loads. These roads are not presently feasible for HW transportation. Since inner route segments do not differ greatly in terms of accident release probability (based on assumed paved conditions for currently infeasible routes), then the most cost-effective alternative is to utilize already paved roads. Appropriate upgrading to a road standard used for heavy trucks is required of roads to be used for HW shipments. Pull-offs and parking spaces are suggested for malfunctioning HW vehicles along these roads, for example.

4. Consideration of Sensitive Populations and Special Public Use Facilities. The sensitive areas analysis identified areas in need of special attention to protect transient or temporary populations along routes over which HW are transported. Such facilities include schools, commercial centers, and stadiums where the population is concentrated. Mitigation measures should be undertaken for these areas including such measures as traffic slow-downs, road quality enhancements, and the scheduling of HW shipments. However, because the release probabilities are extremely low (and even lower when the risk to a specific site is considered), we do not recommend consideration of any zoning adjustments to current land use practices along the routes. The comparatively low levels of release risks coupled with low population densities along

the routes do not warrant adjustments to the present land use patterns.

Preparedness Enhancement. Once a HW incident has occurred, the ability to limit the severity of its effects on public health and safety relies on response capabilities. Preparedness enhancement measures include acquisition of appropriate equipment (including communications equipment), personnel training, and evacuation planning.

The appropriate state and local agencies should continue to work to develop cooperative agreements and mutual aid agreements with the goal of ensuring that fire departments at key points along the transit routes would be adequately equipped and prepared to respond to a HW incident. Intergovernmental agreements among cities and the counties are needed to improve their capabilities in hazardous chemicals management. Despite improvements made over the last three years, there is a need to integrate response efforts, especially when interjurisdictional events occur (e.g., those which could involve the GRIC police and the county sheriff or Department of Public Safety officers). Insufficient funding for equipment attainment and personnel training continues to be a major reason for a lack of expertise, especially among volunteer fire personnel.

References

F. G. Bercha and Associates. 1980. Risks associated with the transportation to treatment of hazardous waste substances: Phase I. Environmental Council of Alberta.

IDF, Inc. 1984. Assessing the releases and costs associated with truck transport of hazardous wastes. Environmental Protection Agency.

Pijawka, K. D., A. E. Radwan, and J. A. Soesilo. 1986. Transportation of hazardous materials in Arizona. Arizona Department of Transportation, Phoenix, AZ.

Rhoads, R. E., et al. 1978. An assessment of the risk of transporting gasoline by truck. Batelle Pacific Northwest Laboratory for the U.S. Department of Energy. November.

SOCIETAL-INDIVIDUAL RISKS FOR HAZMAT TRANSPORT

F.F. Saccomanno and J.H. Shortreed[1]

ABSTRACT

Risks associated with the transport of hazardous materials by truck and rail are viewed from two perspectives: society in general and the individual residing adjacent to the route. A range of societal and individual risks are estimated for the bulk transport of liquefied chlorine gas along a typical transport corridor. Differences in societal and individual risks are assessed for both truck and rail transport options.

The results of this analysis suggest that individual risks associated with the bulk transport of chlorine gas by truck and rail are low, and in the acceptable range. These individual risks were also found to vary significantly between the two modes. Societal risks are much more significant than individual risks, given their concern with very low frequency - high consequence events. For the bulk transport of chlorine along the selected corridor, societal risks for trucks are moderately higher than for rail. Consideration of both individual and societal risks renders the risk analysis process more complete for the purpose of decision-making.

INTRODUCTION

Objective risks associated with the transport of hazardous materials can be viewed from two perspectives: risks to society and risks to individuals at designated locations along a given transportation corridor.

---
[1] Institute for Risk Research, University of Waterloo, Waterloo, Ontario, Canada, N2L 3G1.

Concerns voiced by individuals may differ significantly from concerns voiced by society in general, and this raises serious questions as to the efficacy and practicability of safety policies where only one of these points of view has been considered.

If current policies directed at the transport of hazardous materials are to be justified both from a perspective of improved safety as well as financial responsibility, it is necessary that both societal and individual risks be incorporated into the analysis process.

Societal risks reflect the expectation of all possible damages posed by an activity over an extended period of time, for all locations that are likely to be affected adversely by the activity. For the transport of hazardous materials, these risks are expressed as the product of the probability of a given event and its consequent damage, averaged over all possible consequent damages. In estimating societal risks the full spectrum of consequent damages is considered, from events that are likely to take place normally every year to events that are unlikely to take place even once in one thousand years or more. Furthermore, in estimating the mean societal risk (expected risk), all individuals who may be adversely affected by a given incident are considered, regardless of their distance from the incident or the their likelihood of being affected.

Individual risks, on the other hand, are concerned with potential threats to individuals residing at specific distances from a given transportation corridor or route. These location-specific risks are normally expressed in terms of the probability of death per year or the loss of life expectancy.

The objective of this paper is to compare societal and individual risks for the bulk transport of pressure liquefied chlorine gas by truck and rail tanker. For illustrative purposes, the risks considered in this analysis are estimated for road and track conditions found along the Sarnia-to-Toronto corridor in Southwestern Ontario.

A comment is warranted on the selection of

chlorine as a base material for the purpose of risk comparison. The Toronto Area Rail Transportation of Dangerous Goods Task Force (1989) investigated risks from the movement of various hazardous materials by rail, and found that 28% of fatalities estimated were for chlorine-type products, 42% for gasoline, 26% for LPG and 4% for other products. Chlorine was the only product that posed significant risks at appreciable distances from the rail line. For individuals, risk levels extend up to 7 kilometres from the route. For society in general, the worst case event could possibly result in many thousands of fatalities. While chlorine is not normally transported in bulk by truck, it represents the extremes of risk for hazardous materials in general from both a societal and individual perspective, and serves as a good basis for comparing risk measures for road and rail transport. For this comparative analysis of transport-related risks, the choice of chlorine is also appropriate, since it has the most impact on individuals residing at greater distances from the transportation route.

ESTIMATION OF SOCIETAL AND INDIVIDUAL RISKS

This section describes the various components of the risk analysis process, as applied to the transport of chlorine in bulk by truck and rail tankers.

Risk Components

The estimation of risk is based on two constituent dimensions: probability and consequence. The probability dimension includes the occurrence of an undesirable event (eg. a train derailment or truck accident involving dangerous goods), and the conditional probability that this accident/event will result in a release of hazardous material for different release rates and volumes. The consequences of these releases depend on the hazard areas associated with each likely release profile and on the number of people affected within the corresponding hazard areas for different classes of damage (eg. fatalities, personal injuries, property damage).

Factors that affect the extent and likelihood of accident-induced releases depend on the physical properties of the rail/truck tanker containment system, the operating speed of the vehicle at the time of the accident, and the distribution of forces generated by the accident. The release mode (nature of component release) affects both the rate and volume of material included in each spill. Control factors on the consequent damages (eg. number of fatalities) include the spill environment, material properties and the distribution of population in the vicinity of each incident. Various hazard areas can be determined for each damage level under consideration. Each of these areas has associated with it a term for the probability of occurrence, that is based on the accident/release likelihoods for different release volumes and rates.

Societal risks are obtained by multiplying the accident and release probabilities for different release rates and volumes, by the associated hazard areas and the number of people affected for all classes of damage. As such, societal risks for chlorine reflect an average of all possible damages posed by an activity for all possible hazard area configurations. In the case of chlorine, this refers to all possible configurations of plume dispersal for a given release, as these configurations are affected by changes in atmospheric stability and changes in wind directions and speeds. Individual risks treat the accident/release probabilities separately from a designated consequent damage. In this latter measure, probability and consequences are not multiplied together to yield a single measure of expectation, but are considered separately as a relationship between distance from a given incident along a right-of-way and the probability of damage for a predetermined period of time and an assumed shipment volume. For a corridor analysis, individual risks tend to depend mainly on distance from right-of-way, accident involvement and conditional probabilities of releases in an accident sitiuation. Societal risks, on the other hand, tend to vary with population distributions and densities, and land development patterns present along various corridor sections.

Typical Societal and Individual Risk Estimates

In this section, various components of risk are estimated for the road and rail transport of liquefied chlorine gas. These estimates are generic in nature, in that they can be applied to any road/rail corridor by adjusting for the appropriate population distribution.

Releases from containment can take place under two conditions: normal transport incidents and as a result of an accident. Normal incidents, while taking place with some frequency, generally involve a small amount of material resulting in low, confined consequent damages. "Normal" releases are caused by either tanker valve leakages or defects in the tanker welds. Accident-induced releases, on the other hand, take place with much lower frequency but reflect more catastrophic failures and more extensive consequent damages. In this analysis, we will be concerned solely with these latter accident-induced releases.

Aggregate release probabilities were estimated for each road and rail accident using a mechanistic, fault tree analysis of the tanker containment system (Saccomanno, Shortreed and Thomson, 1990). For chlorine bulk tankers, two containment system faults were considered: releases from tank shells (includes tank wall, tank head and manway cover failures) and releases from valves (includes pressure relief valves, liquid valves, outlet valves etc.)

Table 1 summarizes the major tanker release modes (component failures) for rail and road shipments, respectively. These tanker fault probabilities suggest that for chlorine rail shipments, approximately 1.5% of all railcar accidents result in a loss of lading. This value ignores the frequent venting of pressure relief valves in an accident, that often results in a minor loss of material. Consideration of releases through pressure relief venting increases the conditional release probability to over 10% of all accident occurrences. Release faults for rail tankers are compared in Table 1 with similar fault occurrences for truck transport. For trucks approximately 1.6% of accidents produce a loss of lading. The fault tree analysis which produced these estimates was based on failure structures developed by Batelle (Andrews, et.al., 1980), and modified to reflect Canadian conditions by the Institute for Risk Research (IRR, 1987).

Conditional fault probabilities estimated from fault trees appear to be under-estimating values quoted

Table 1 * Accident - induced release probabilties for different tanker faults (release per 100 tankcar accidents).

	Rail	Road
Tank Shell	1.097	1.460
Tank Head	0.067	..
Tank Wall	0.995	..
Manway Cover	0.035	..
Tank Valves	0.453 **	0.165
Gas Valve	0.165	..
Relief Valve	(10.200)	..
Liquid Valve	0.165	..
Outlet Valve	0.123	

* output from fault tree analysis
** excludes normal pressure valve venting

elsewhere in the literature. Based on the US experience, Harwood (et.al. 1989) has reported values in the range 5-10% for accident-induced releases of hazardous materials across various road types. Comparisons with limited release data from Canada for 1985-86 (CANUTEC, 1989) and ignoring minor relief valve release, however, appear to support lower rates in the range 1-5% of hazardous material accidents. These differences may be due to the fact that the Canadian data are more recent in nature, reflecting major improvements in the last few years regarding tank car design and operations.

Among other factors, the hazard area associated with each incident involving road and rail shipments of chlorine is affected by the type and volume of material released. In this study, two types of release have been considered: instantaneous and continuous releases. Instantaneous releases refer to a situation where the bulk of material is released immediately following an accident. Continuous releases, on the other hand, take place over an extended period of time, in some cases up to several hours. Under certain conditions, some releases can be both instantaneous and continuous. For each type of release, three volume/rate classes have been considered for both rail and truck modes:

Instantaneous Releases

Tanker Volume	
High	100%
Medium	69%
Low	39%

Continuous Releases

Rate kg/s	
High	14.5
Medium	3.9
Low	0.1

Before obtaining estimates of release frequencies for different volumes and rates of material involved, it was necessary to establish a relationship between the containment system failure mode (from the fault tree analysis) and the volume and rate of material released. In order to determine the conditional probabilities of release for different failure modes, all incidents involving compressed gases (Class 2 incidents) as reported in the CANUTEC data file were analyzed for the 1986-87 period (IRR, 1987). There were a total of 38 such releases observed, of which 6 took place on rail and

32 on road. These releases were grouped according to the type and size of release and by the primary containment system failure mode (ie. shell or valve). All major spills were assumed to be instantaneous in nature, while minor leaks were assumed to be continuous. In addition, all valve releases were assumed to be continuous. Some releases reported in the CANUTEC data base could not be used in this analysis, since they were not identified as being either shell or valve initiated.

Given the lack of observations in the CANUTEC data base, a significant degree of "intuitive adjustment and smoothing" was applied to the resultant contingency table of releases to reduce the number of empty cells. Table 2 summarizes the proportion of releases associated with different failure modes (shell and valves) for different release profiles (instantaneous and continuous). These proportionalities were estimated for road and rail shipments involving all hazardous materials, and have been applied here to bulk chlorine shipments. Hazardous materials as used in this analysis include all materials classed as " Special Dangerous Goods" under 1985 Transport Canada regulations (Institute for Risk Research, IRR, 1987).

Combining the release probabilities from Table 1 with the proportions from Table 2 yields the conditional accident-induced release probabilities for different release types and rates (instantaneous and continuous modes). These values are summarized in Table 3 for road and rail.

Considering the consequence side of risk, the nature of the release affects the resultant hazard area and the damages corresponding to these areas for a given location. All other factors assumed constant, the potential for damage within these hazard areas is affected by the distance from each release, ie. the closer to the release the greater is the potential for damage. The result of this analysis produces individual risk isopleths, that represent distances of equal damage probabilities for different types of damage . This reflects the individual risks associated with the activity under consideration. When these individual risks are summed over all distances from a given release, the resultant measure reflect

Table 2. Proportionalities of material released by type and containment fault (Based on CANUTEC Data).

	Release Proportionalities (%)					
	Instantaneous			Continuous		
	High	Medium	Low	High	Medium	Low
Fault Type						
Rail						
Tank Shell	20.0	40.0	35.0	5.0	----	----
Valves	----	----	----	25.0	35.0	40.0
Road						
Tank Shell	20.0	25.0	35.0	10.0	5.0	5.0
Valves	----	----	----	30.0	30.0	40.0

Source: Saccomanno (et.al.,1990)

Table 3. Tankcar release probabilities ($\times 10^{-3}$) in an accident situation (Chlorine).

Rail

	High	Medium	Low
Instantaneous	2.19	4.39	3.84
Continuous	1.63 (27.18)*	1.59 (37.29)	1.81 (42.61)

Road

	High	Medium	Low
Instantaneous	2.92	3.65	5.11
Continuous	1.96	1.23	1.39

* value in brackets includes pressure relief valve venting

Source: Saccomanno (et.al.,19

the corresponding risks to society for each incident.

The hazardous areas associated with chlorine atmospheric dispersion have been estimated in this study, using a Gaussian dispersal model developed by Homann and Associates (EPI, 1989). The results of the EPI model were compared for consistency with hazard areas estimated by the Institute for Risk Research (1987), and were found to be in general agreement.

Critical distances from each incident depend on the assumed level of damage associated with each release. In this analysis, two classes of fatality impact have been considered: 50% fatality and 1% fatality. For chlorine, these 50% and 1% fatality isolines correspond to critical concentration levels of 800 and 300 ppm, respectively. The percentage in these criteria refers to the proportion of people killed within a given critical distance from each incident, or the probability that an individual will be killed given the level of exposure to a chlorine plume at the specified concentration.

The hazard areas associated with the 50% and 1% fatality isopleths are summarized in Tables 4 and 5 for rail and road bulk shipments of chlorine, respectively. To obtain a measure of societal risk, that is independent of population distributions along a given transport corridor, a weighted average fatality rate per unit density was estimated for each hazard area using the following expression:

$$FR = [HA_{1\%} - HA_{50\%}] * KR_{30\%} * SF + HA_{50\%} * KR_{80\%} * SF \quad (1)$$

where

FR = deaths per capita of exposed population in the hazard area.

$HA_{1\%}$, $HA_{50\%}$ = hazard areas for 1% and 50% lethality.

$KR_{30\%}$, $KR_{80\%}$ = average 30% and 80% kill rates.

SF = shield factor for people being indoors or who are evacuated in time from the hazard area.

The shield factor in Eq. 1 is based on

Table 4. Hazard areas for different Chlorine release profiles on rail (EPI model results).

Type of Release		Hazard Area 50% Fatality (800 PPM)	1% Fatality (300 PPM)	Fatalities** per density exposed
Instantaneous				
	High	0.971	2.157	0.113
	Medium	0.759	1.693	0.089
	Low	0.544	1.191	0.063
Continuous				
	High	0.113	0.384	0.017
	Medium	0.025	0.094	0.004
	Low	0.0001	0.001	0.00004

** Fatality rates based on a 1 person per sq. km. density within the hazard area.

Assumed wind speed 5 m/s

Atmospheric stability condition D.

Table 5. Hazard areas for different Chlorine release profiles on road (EPI model results).

Type of Release		Hazard Area 50% Fatality (800 PPM)	1% Fatality (300 PPM)	Fatalities per density exposed
Instantaneous				
	High	0.472	1.020	0.054
	Medium	0.395	0.856	0.045
	Low	0.314	0.675	0.036
Continuous				
	High	0.113	0.384	0.017
	Medium	0.025	0.094	0.004
	Low	0.0001	0.001	0.00004

** Population density of 1 pers. per sq. km.

Assumed wind speed 5 m/s.

Atmospheric stability condition D

air infiltration rates for typical Canadian houses assuming that 90% of the population are indoors at the time of the incident or can be evacuated from the scene in advance of the toxic plume. An in-depth analysis of air infiltration properties was not carried out for this study. Such an analysis may be found in Wilson (1989) for Canadian conditions and in Purdy (1988) for the UK. In this analysis, a shielding factor (SF) of 0.10 was applied to both road and rail incidents involving chlorine. This factor assumes that only 10% of expected fatalities from the damage propagation models would actually be killed, since most people would be indoors during the incident, and hence be shielded from the full impact of the hazard. The 10% factor also reflects evacuation possibilities in advance of the plume dispersal. The actual numbers of people killed given an incident involving chlorine can be obtained directly by multiplying each of the fatality rates in Tables 4 and 5 by the corresponding population densities in the vicinity of each spill.

RESULTS OF SARNIA - TORONTO CORRIDOR ANALYSIS OF RISKS

For the purpose of analysis, it is assumed that 900,000 tonnes of chlorine are shipped annually along the 200 km Sarnia-Toronto rail/road corridor for each of the two modes under consideration. For an assumed payload of 90 tonnes per rail tanker and 27 tonnes per truck tanker, the total annual shipment of chlorine corresponds to 10,000 railcar and 33,300 truck movements along the corridor, for a total exposure of 180 million tonne-km annually on each of the two modes. It should be noted that the societal and individual risk values estimated for this exposure can be adjusted to include any change in annual shipments on each mode over any section of the corridor.

Average accident rates were estimated for the Sarnia - Toronto corridor, based on rail and truck accident data for the period 1980-85. (IRR, 1987). These rates are 0.0052 railcar accidents per million tonne-kms for rail and 0.036 truck accidents per million tonne-kms for road. Minor variations in accident rates were obtained for various sections of the Sarnia-Toronto corridor for both rail and road sections (Saccomanno, Shortreed and Van Aerde, 1989). These section-specific rates can be applied directly to the corridor analysis, in lieu of using a corridor-wide average rate. This approximation was not deemed to affect the results appreciably, such that a section-specific application of accident rates for road and rail was not done for this paper.

Societal Risk Estimates

Since societal risks are estimated over an extensive area of impact, it is necessary to input information on population distribution along various sections of the corridor. In this analysis, an average population density of 600 persons per sq. km. has been assumed for the corridor, based on the actual rail route densities for a mix of urban and rural sections along the right-of-way. Applying this average population density to the hazard areas in Tables 4 and 5 yields cumulative F - N plots of annual fatalities versus their corresponding probability of occurrence for chlorine for each of the two modes under consideration (Figure 1). The annual probabilities per consequent fatality in these F - N plots were obtained by multiplying the accident-induced release probabilities from Table 3 by the corresponding average accident rates per tonne-km for rail and road, and the annual tonnage assumed for the Sarnia-Toronto corridor. Each probability terms corresponds to a specific release type, ie. instantaneous/continuous and high/medium/low.

The resultant F - N curves (Figure 1) for rail and truck transport of chlorine are downward sloping. As the number of fatalities (N) increases, the cumulative frequency of release required to sustain this damage (F) decreases rapidly. A comparison of F - N curves between road and rail yields some interesting results. For chlorine, differences in the F - N curves for road and rail transport are not very pronounced, especially in the lower range of damage. The mean of each F-N curve corresponds to the expected number of fatalities per year along the corridor. For an assumed annual shipment exposure of 180 million-tonne km on truck and rail, our analysis of societal risks along the corridor suggests annual fatality rates of 0.51 and 2.12 deaths per year for rail and truck transport, respectively. Higher risks for trucks are a result of significantly higher accident rates than rail for the same level of exposure. Rail reflects higher hazard

Figure 1. F-N societal risk curves for chlorine transported by rail and truck.

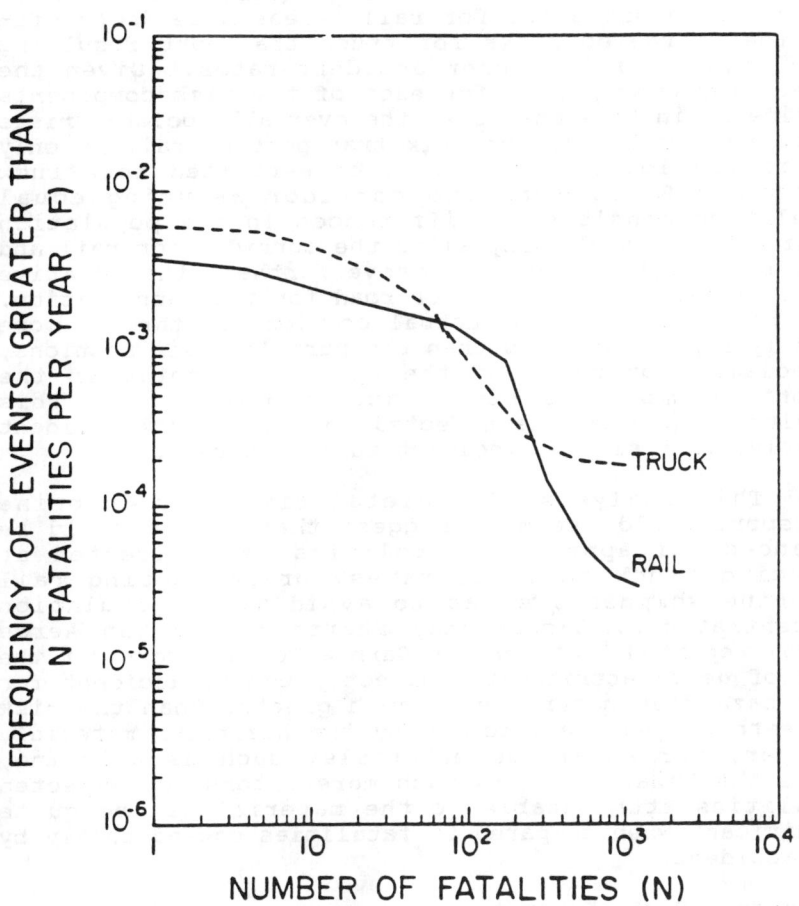

areas than road, since more material is assumed to be carried by each rail tanker and since the dispersal relationship for gases such as chlorine are a non-linear function of the material involved (IRR, 1987). More extensive hazard areas for rail releases tends to off-set the increased risks for truck transport resulting from significantly higher accident rates. Given the underlying assumptions for each of the risk components considered in this analysis, the over-all societal risks associated with chlorine bulk transport by rail is only moderately lower than the risks estimated for truck along the Sarnia-Toronto corridor assuming equal population densities. Differences in the population distribution and density along the corridor for rail and road route options would improve further the relative safety advantage of rail over road for this application. Road sections along the actual corridor go through more densely populated areas than the parallel rail sections, especially for parts of the corridor located in the Toronto Region. All other factors being equal, this results in more people affected for each truck incident relative to a similar incident taking on rail.

This analysis of societal risks for chlorine transport would seem to suggest that safety could be enhanced, if appropriate policies were directed at reducing truck accident rates, or re-routing rail chlorine shipments so as to avoid major population concentrations. Saccomanno, Shortreed and Van Aerde (1989) reported that for the Sarnia-Toronto corridor the risk of death attributable directly to the accident for most hazardous materials is much greater than the risk of death caused specifically by the hazardous material. However, for hazardous materials, such as chlorine, where the hazard area is much more extensive, expected fatalities attributable to the material can be quite significant when compared to fatalities caused solely by the accident.

Individual Risks Estimates

Individual risks were calculated for chlorine by considering eight different kinds of weather conditions for each spill environment. These include atmospheric stabilities: F, B, and D, and wind speeds of 2, 5 and 8 meters per second (8 not used with F). The percentage distribution of wind speeds and atmospheric stability were taken from the Toronto

International Airport weather station. It was determined that there were minor differences in individual risk contours when wind direction was considered (Concorde, 1988). For this analysis, winds were assumed to be equally likely from all directions.

For each spill type given in Tables 4 and 5 (rail and truck modes, respectively), plume centreline distances were obtained using the EPI (1988) chlorine dispersal relationships for the 50% and 1% fatality concentrations. For 65% of instantaneous releases, it was assumed that none of the liquid chlorine was entrained in the release, while for the other 35% of the events it was assumed that 100% of the tank contents were released. Furthermore, a standard flat level topography was considered for continuous releases, while for instantaneous releases, an urban topography was considered in the dispersal relationship.

Individual risk is clearly higher for high volume locations and lower for low volume locations, and must be revised according to the volume of traffic. In this analysis, individual risk estimates were expressed in terms of the annual probability of death at specific distances from the rail/road route section. The annual probabilities are based on an assumed corridor shipment volume of 180 million tonne-kms. per year for each mode, with all individuals being exposed to this traffic for the entire year.

Since individual risks are location-specific, overlaps in hazard areas for various releases profiles need to be considered explicitly in the analysis. An individual residing at a specific point in space may be affected by any number of release profiles at various spill locations along the route. Overlaps in hazard areas at specific locations can occur as a result of two conditions: changes in wind direction and speed affecting the plume configuration downstream from a given release, and movement of the source of release, ie, the vehicle, along a given route section.

To calculate the individual risk at some distance (d) from the route, the source of a spill must be moved along the route for each circular hazard area (reflecting all wind directions). Figure 2 illustrates the method used to calculate the exposure to an event at a distance d = 0.2 km from the route, assuming D5 weather. For the 50% and 1% fatality isopleths, the exposure adjustment was taken as the cord length (in kilometres) on the hazard area circle at a distance d from the centre. This cord length is equal to the distance along the route for which events would impact a given point in space for a designated plume concentration level. Since the accident rate is expressed per tonne-kilometer, the length of the cord in kilometres represents the probability of occurrence of events that would impact a given point adjusted by the vehicle-distance traversed along the route.

The estimated individual risks for road and rail are presented in Figures 3 and 4. The probabilities for the 50% and 1% concentration levels are given for each mode based on the movement of 900,000 tonnes along the route and the given accident rates per tonne-km for each mode.

In estimating these probabilities, the fatality areas were assumed to have a triangular shape with a cone angle of 20 degrees. Thus for the full 360 degree circle, the event probabilities were reduced by a factor of 20/360 or 1/18 to account for variation in the wind direction. Thus the assumption of equal wind directions leads to a relatively simple approach for calculating individual risks at a given distance from the route.

Since truck accident rates are significantly higher than rail accident rates, trucks were found to reflect higher individual risks nearer to the route. On the other hand, since rail tanker cars are assumed to carry more material, rail risks for chlorine transport were found to be higher than truck risks at greater distances from the route. More material per tanker means that rail is subject to more extensive hazard areas than truck given a release of hazardous material.

The individual risks for either truck or rail are reasonably low using typical values of 10^{-5} to 10^{-6} as indicative of "de minimus" risk for an individual (Figure 4). At 10^{-5} the statistical risk is equivalent to the loss of about 10 days of life and at 10^{-6}, 1 day of life. Risks associated with the

Figure 2. Adjusting risk exposure at distances from route by variation in plume configuration and moving release source (D5 Weather).

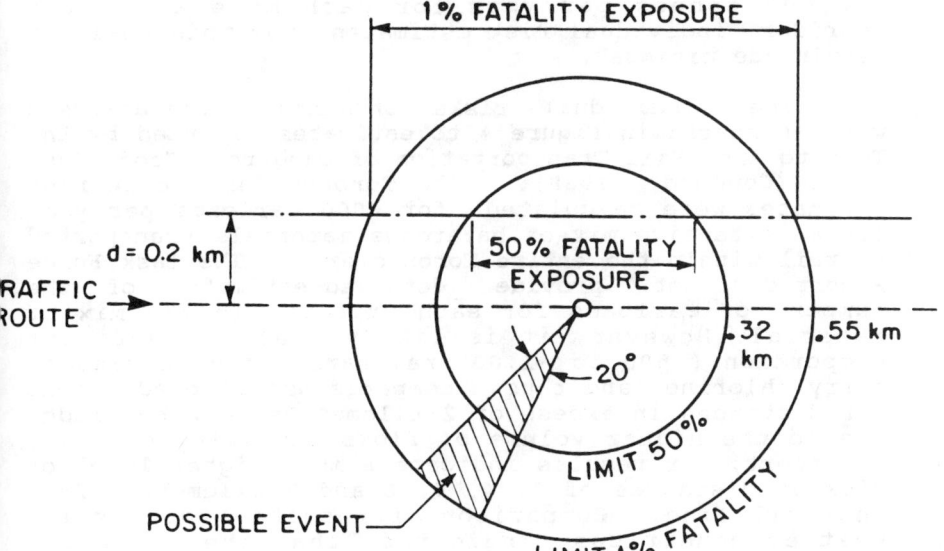

transport of chlorine by truck and rail are in the range of "acceptable risks", ie, less than 10^{-6} for both modes, which can be compared with a 10^{-5} chance of being struck by lightening. It should be noted that these risks vary directly with the volume of traffic along the route, so with 1,000 chlorine rail cars per year, all the individual risks would be lowered by a factor of 10. Even assuming a relatively high volume of 900,000 tonnes per year for each mode along the corridor, individual risk estimates from this analysis remain "de minimus".

The individual risks obtained in the analysis were compared in Figure 4 to estimates reported by the Toronto Area Rail Transportation of Dangerous Goods Task Force (Concord, 1988). The Toronto Task Force risk estimates were calculated for 6000 carloads per year for an extensive mix of hazardous materials transported by rail within the entire Toronto area. The Task Force Report did not provide detailed estimates of the number of carloads for each product in the mix of material. However, it is likely that a signficant proportion (500 to 1,000 railcars) were assumed to carry chlorine, and thus comparisons of predictions for distances in excess of 2 kilometres can be made. Due to the higher volume of flows involving chlorine transport, our results indicate a much higher level of risk at distances of 3, 4, 5, 6 and 7 kilometres from the rail line. Comparison of results for similar weather conditions indicated that the EPI model predicts 1% fatality level concentrations at distances as much as 50% higher than results reported by the Toronto Rail Task Force.

The differences between the Toronto Rail Task Force results and the results of our analysis in Figure 4 are representative of the impact of underlying assumptions on suggested risk levels for the transport of hazardous materials. For example, the choice of material, weather conditions, release events, nature of the plume dispersal expression, etc. can each affect significant differences in risk prediction. Irrespective of underlying assumptions, however, the results are consistent in suggesting that individual risks associated with the transport of hazardous materials by rail and truck are negligible at various distances from the route.

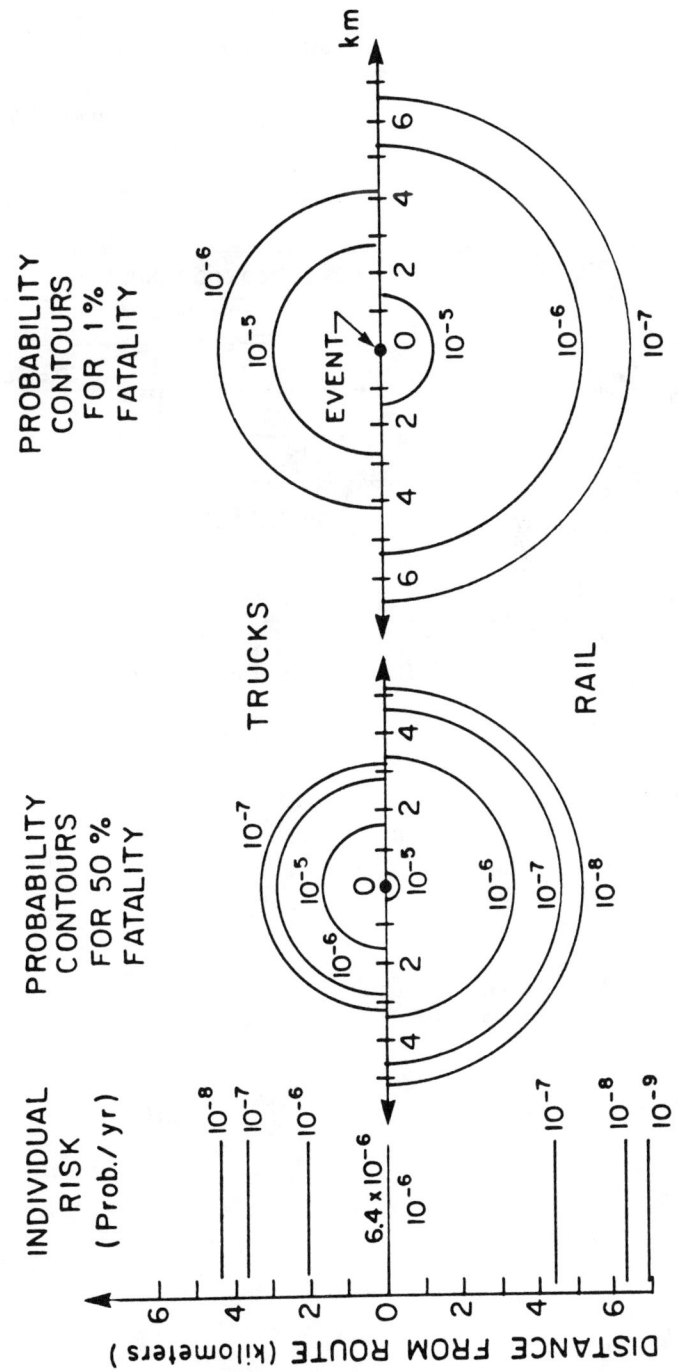

Figure 3. Individual risk isopleths for rail and truck (900,000 tonnes of chlorine per year).

HAZARDOUS MATERIALS TRANSPORTATION

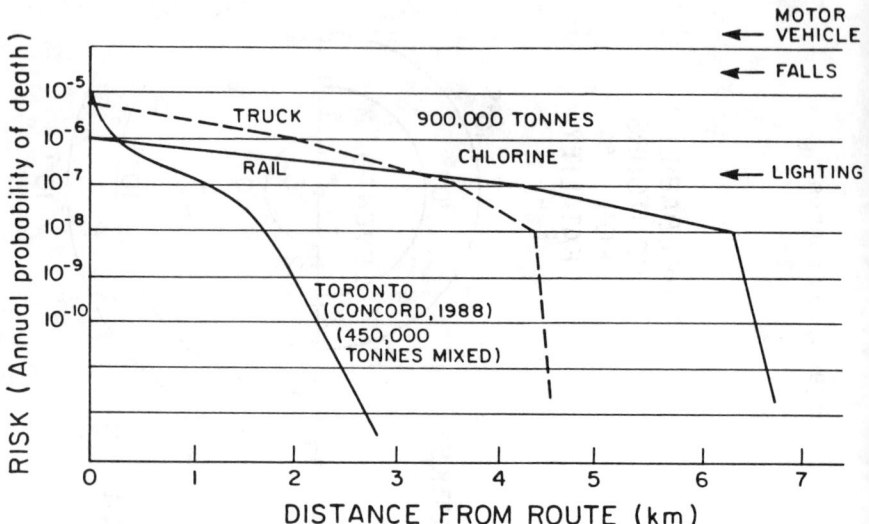

Figure 4. Individual risk versus distance from route for rail and truck (900,000 tonnes of chlorine per year).

CONCLUSIONS

For a relatively high chlorine shipment of 900,000 tonnes, both society and individual risks were found to be low. For both truck and rail transport, individual risks were less than 10^{-5} per year for all locations, and below 10^{-6} for most locations adjacent to the route. These results suggest that, from an individual risk perspective, we need not be too concerned with the transport of hazardous materials by rail and truck. The risk of death from other causes (eg. being struck by lightening) appears to be decidedly more acute. On the other hand, for our assumed 200 km corridor with 900,000 tonne throughput and a 600 person per square Km density, the societal risks were found to be significant and warrant some attention (ie. 0.51 fatalities per year for rail and 2.12 fatalities per year for road).

For distances extending to 2 1/2 kilometres from the route, the individual risks for chlorine were found to be higher for trucks than for rail, mainly due to the higher truck accident rates per tonne-kilometer. Beyond this distance the individual risks were found to be higher for rail, mainly due to the larger quantity of material carried per tanker.

Individual risk is signficantly responsive to changes in the volume of traffic on the route and distance from the route. In addition to volume, societal risks also vary with population distribution and densities along the route.

When evaluating the risks of transporting hazardous materials by rail or truck, both individual and societal risks need to be considered. Individual risks reflect the level of concern to persons living and working adjacent to the route. Since individual risks are not summed over all distances from the route, they are orders of magnitude lower than societal risks for the same material in transit.

The order of consideration for individual and societal risks is important. Since individual risks best reflect the concerns of adjacent residents and workers, and since they are traditionally the first consideration in risk assessment, it is recommended that individual risks be considered first in the evaluation process. If these risks are unacceptable, then is not necessary to even evaluate the broader societal risks. If individual risks are acceptable, then it would be necessary to consider the larger societal risks, particularly as they relate to the occurrence of very low frequency- very high conseqence events. In general, the communication of these individual risks and their relatively low levels should lead to more informed decision-making.

REFERENCES

1. Andrews, W.B. et.al. (1980). An Assessment of Risk of Transporting Liquid Chlorine by Rail. PNL-3376. Pacific Northwest Laboratory. Richland. WA.

2. Canadine, I.C. and Purdy, G. (1988). The Transport of Chlorine by Road and Rail in Britain - A Consideration of Risks. Paper presented to the 6th International Symposium of Loss Prevention in the Process Industries. Oslo, Norway. June 19-22nd.

3. CANUTEC. (1989). Canadian Transport Emergency Centre. Reported Releases of Dangerous Goods During Transport by Truck and Rail. Transport Canada.

4. Concorde Scientific (1989). Consultants Report to the Toronto Area Rail Transportation of Dangerous Goods Task Force.

5. Harwood, D.W., Viner, J.G. and Russel, E.R. (1989). A Truck Accident Rate Model for Hazardous Material Routing. Presented at the 1989 Annual Meeting of the Transportation Research Board. Paper No. 890458. Washington, D.C. January.

6. Homann and Associates. (1989) Users Manual for the EPI Model.

7. Institute for Risk Research (1987). Risk Analysis for the Handling and Transportation of Dangerous Goods. Final Report to the Transportation of Dangerous Goods Directorate. Transport Canada. December.

8. Saccomanno, F.F., Shortreed, J.H. and Van Aerde, M. (1988) Assessing the Risks of Transporting Dangerous Goods by Truck and Rail. Final Report No. 2. Institute for Risk Research. WRI Award No.

1254501. July.

9. Saccomanno, F.F., Shortreed, J.H., Thomson, M. (1990). Fault Tree Analysis of Releases of Pressure Liquefied Gases from In-Transit Rail Bulk Tankers. Forthcoming in Forensic Engineering. Vol. 2, No. 3.

10. Toronto Area Rail Transportation of Dangerous Goods Task Force. (1989). Final Report for Transport Canada. Supply and Services. Ottawa.

11. Wilson, D.J. (1987). Stay Indoors or Evacuate to Avoid Exposure to Toxic Gas? Emergency Preparedness Digest. Vol 14. No. 1. January-March. pp. 19-24.

RISK MANAGEMENT AND ADVANCED TECHNOLOGIES

GIS RISK ANALYSIS OF HAZARDOUS MATERIALS TRANSPORT

Charles Anders[1] and James Olsten[2]

Abstract

The Geographic Information System (GIS) was used to assess the risks and vulnerability of transporting hazardous materials and wastes on the Arizona highway system. This paper discusses the methodology that was utilized, and the application of GIS systems to risk analysis problems.

Introduction

The transportation of hazardous materials and wastes has become a routine part of our society. Past studies have shown that one in ten trucks traveling on our highway system is transporting some type of material that is classified as hazardous. This fact presents a challenge to those government officials and professionals in the private sector that are responsible for the safety of the public.

In January of 1986, a study was published which estimated the magnitude, chemical types and hazard class of hazardous materials being transported on the Arizona highway system. This study was conducted by the Center for Environmental Studies, the School of Public Affairs and the Center for Advanced Research in Transportation at Arizona State University in Tempe, Arizona. The study was funded by the Arizona Department of Transportation and the Federal Highway Administration and administered by the Arizona Transportation Research Center.

---

[1]Associate, Dames & Moore, 7500 N. Dreamy Draw Drive, Suite 145, Phoenix, Arizona 85020.

[2]Senior Engineer, Dames & Moore, 1125 Seventeenth Street, Suite 1200, Denver, Colorado 80202-2027.

The 1986 study entitled "Transportation of Hazardous Materials in Arizona" utilized a number of surveys to estimate the type and volume of hazardous materials and hazardous waste being transported on over 80 highway segments throughout Arizona. A follow-on study conducted by Dames & Moore was initiated to utilize the data developed in the 1986 study and evaluate the risks associated with the movement of these hazardous materials. In addition, this study assessed the vulnerability of populations in geographical areas by integrating the emergency response times and capabilities into the analysis.

This study was conducted using a geographic information system which is a new approach to risk analysis. Use of the geographic information system provided a more detailed evaluation of how the transportation system interacts with a wide variety of demographic factors, such as population and land use. Additionally, a geographic information system easily allows consideration of specific site features such as locations of emergency response units, hospitals, schools, etc.

This paper will discuss the methodology which was developed to conduct the risk analysis and the application of the GIS to address various issues associated with the transportation of hazardous materials and hazardous wastes.

The application of these analysis techniques are illustrated on the Arizona highway system network. This information is presented for illustration purposed only and no conclusions regarding the actual risks of transporting hazardous materials in Arizona should be drawn from these data.

Data Analysis

A broad range of data were utilized to conduct the risk assessment. The estimated volume of hazardous materials and hazardous wastes that are transported on the state highway system were categorized by hazard type. The amount and number of truck trips of each type of material was estimated for over 80 individual highway route segments. An illustration of the typical volume data is presented in Figure 1.

The truck accident rate for individual highway segments was obtained from the Arizona Department of Transportation. These data were broken down into over

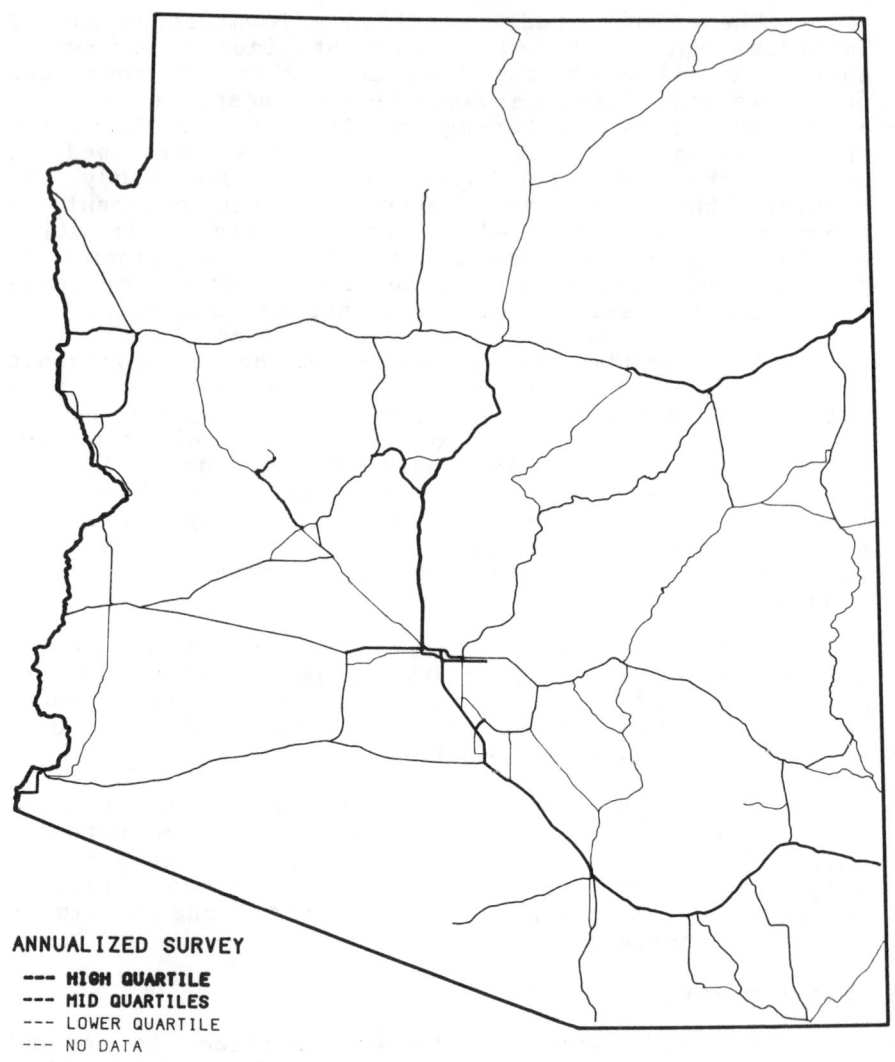

FIGURE 1

600 individual segments. Figure 2 presents typical truck accident rate data.

Population data for the complete State of Arizona were utilized to assess the actual population at risk. These data were obtained from the 1980 census and are presented in Figure 3.

Information on the location of emergency response units was obtained from the Arizona Division of Emergency Preparedness of the Arizona National Guard. This information included the address of each of the fire department station located throughout the State.

Risk Assessment Models

The objective of modeling the risk for this study was to obtain a comparative analysis for the entire state. The spatial relationships were more important than the specific values obtained, that is the relative risks across the network were the principal objective.

Models can be used for varying purposes. For this study, the risk models were used as an analysis tool for understanding the data and allow for planning activities on the network. As such the entire network, was modeled. Alternative models can be described as fixed specific location models and real time event models. Specific location models can typically be calibrated with annual meteorological data incorporating the probability of wind direction and speed. Event models require actual meteorological data and are used in predicting the consequence of an actual event in emergency response functions.

The basic risk model is described in Federal document FHW-1P-80-15, Guidelines for Applying Criteria to Designate Routes for Transporting Hazardous Materials. It consists of:

The frequency of hazardous material shipments

Probability of an event at a location

The nature of dispersion based on meteorological factors

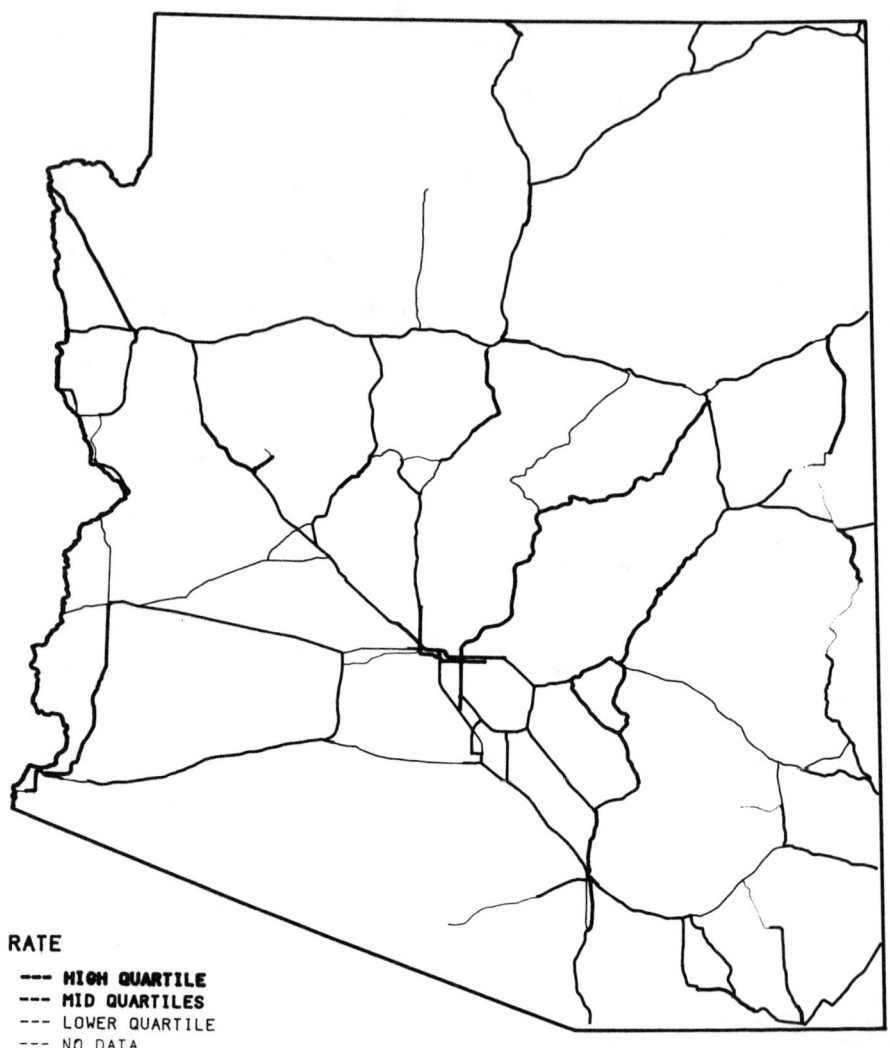

AVERAGE ACCIDENT RATES

FIGURE 2

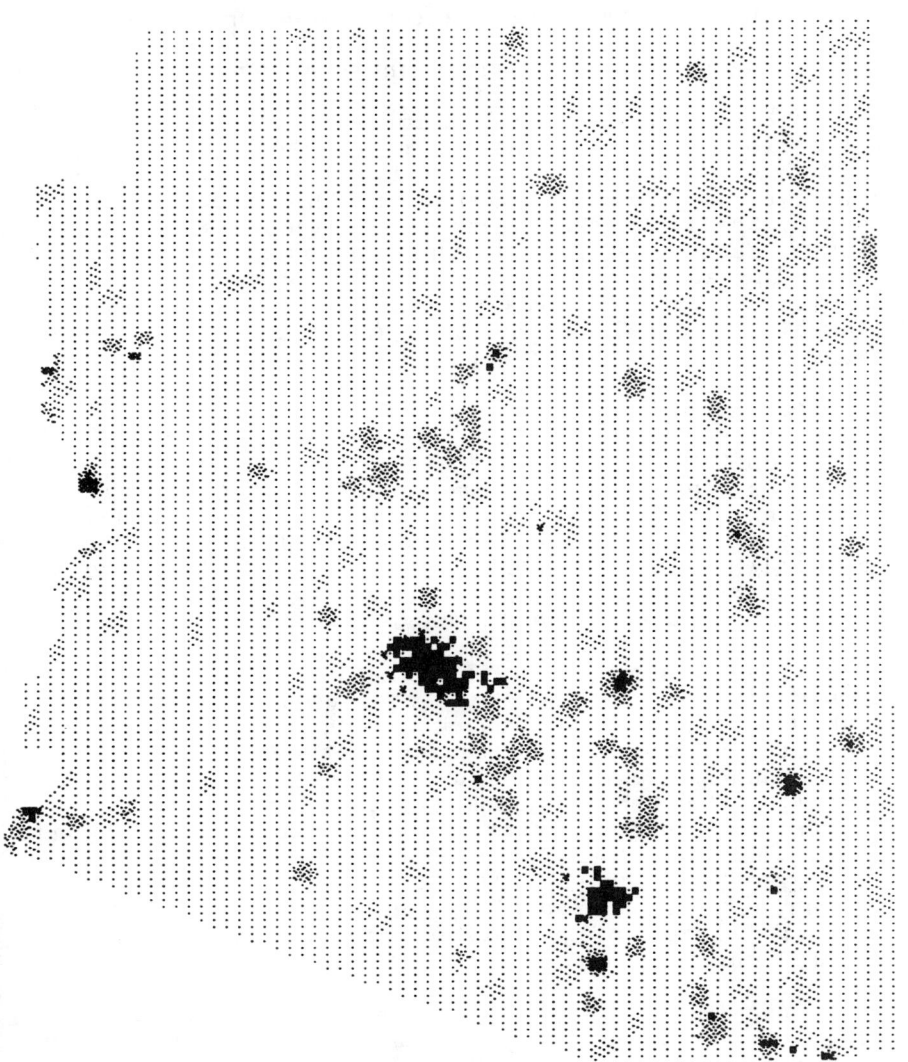

POPULATION

FIGURE 3

The density of population in the area

The nature of the incident consequences

For this study is was impractical to incorporate meteorological conditions for over 100,000 square miles of Arizona. Since the severity of the event is also relative, we chose to externalize this factor from the model itself. The severity consequence is not a spatial factor, while all other components of the model are. The ability to externally weight the final model results allows for varying the consequences based on medical, chemical and public perception of the event type.

Database Methodology

Spatial analysis requires locational diversity. For this project the road network provides the framework for analysis. The federal aided highways in Arizona were digitized based on the USGS state series maps which are in a LAMBERT projection. The state and county boundaries are obtained from digital data prepared for the 1980 census.

The major data integration occurred in a combination of tabular traffic data from state surveys with the geographic network. The digitized network consisted of the unique segments between intersections/interchanges making up the federal aided highway system. The tabular data typically referred to only the Arizona Port of Entry and destination. A database model was specified for each origin/destination which allocated the surveyed trip to the constituent segments in the network. Essentially each survey shipment was projected into unique segment records in the detailed tabular database.

Accident data from the state records were available in more detail than were the trip routes. Within the GIS, the segments were coded in a hierarchy. This hierarchy allowed flexibility to analyze segments as a whole when using the tabular data on materials shipments and also to assess the more detailed accident data on such segments. Typically the accident data was available with subsegments defined by major state and county road intersections/interchanges. Most of the federal aided segments were classified into 3 to 5 subsegments with historic accident rates for truck traffic for each subsegment.

Population data were taken from the 1980 MARF summary tapes. While this data has limited geographic information it was the best available prior to publication of TIGER data from the 1990 census now being processed. Two summary levels in the MARF file contain geographic centroids, these were read into the GIS and aggregated on a square mile basis. Inspection of the maximum values indicated that normalizing was required to achieve actual density. The GIS was requested to distribute the population from its inherent "point" location equally across a 5 square mile area thereby smoothing the data and reaching a calibrated persons per square mile density matching the observed development density. In rural areas when the enumeration districts are well over 5 square miles, this population model was conservative in its overestimation of population density.

Modeling Methodology

The data were entered and in the Geographic Information Management System (GIMS), developed by Dames & Moore. GIMS is a comprehensive vector and raster based system conceived, applied and evolved in over 100 geographic analysis projects during the last 15 years.

Five models were constructed based on the generic evaluation equation of four spatially varying components:

Component	Evaluation
Accident rate	Incident Probability
Shipment Frequency	Hazard
Population Affected	Risk
Response Time	Vulnerability

The generic formula for evaluation consisted of the following:

Absolute hazard = (R) (F)
Population at Risk = (R) (F) (P)
Vulnerability = (R) (F) (P) (T)

Where

R   Accident rate by sub-segment, these data are drawn directly from the accident rate table.

F Shipment frequency by highway segment, these data are drawn from the previous survey data.

P Population affected, determined by the local population density and modified by the material specific effects radius.

T Response time, determined by nearness of fire stations as modeled over the principal highway network.

The initial component of the model provides a comparison of the absolute hazard by route segment which is calculated by multiplying the truck accident rate by the volume of hazardous material transported. This information is presented in Figure 4.

The population affected by a hazardous materials incident is calculated by evaluating the population within an impact area. The impact area varies depending on the type of hazardous material being transported and the exposure pathway. The following pathway mode models were utilized using the GIMS command structure.

Model	Representative Material	FTRB Radius In Miles
Inhale	Non-Flammable Gas	2.0
Blast	Explosives	0.5
Toxic	Poisons	0.3
Contact	Corrosives	0.7
Combust	Gasoline	0.5

The population at risk index is then calculated by multiplying the truck accident rate by the frequency of hazardous material by the population affected. An example of the outputs of the population at risk analyses is presented in Figure 5.

The geographic locations of fire stations were included in the map database. Utilizing the Federal Aided highway network as the principal access from the fire stations to potential incidents, both the network and the intervening space were modeled for probable response time. The highway network was set to support a 50 mph effective speed, urban and rural areas off the principal network were set to an effective 25 mph speed. A base mobilization time of 5 minutes was also assumed. The model allows the user to set any of these values for alternative assessments. Based on the above

GIS RISK ANALYSIS

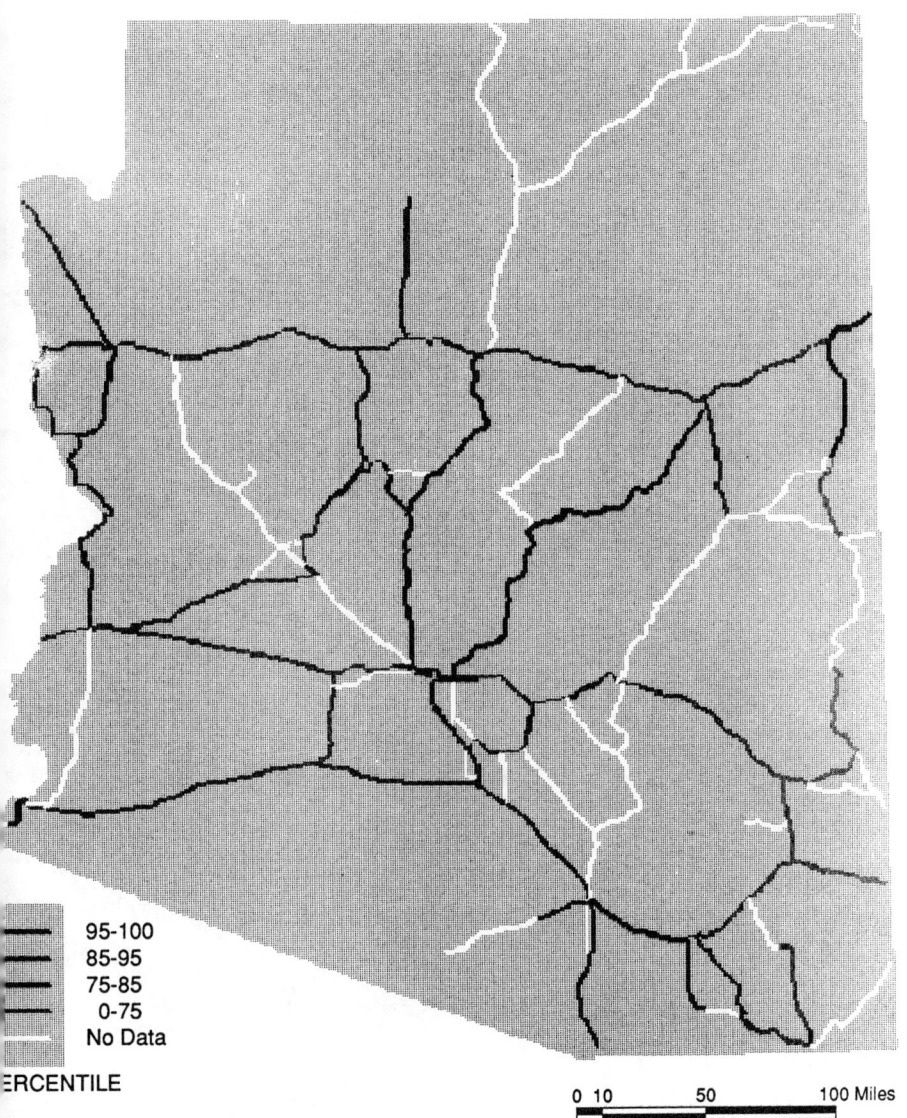

95-100
85-95
75-85
0-75
No Data

PERCENTILE

0 10   50   100 Miles

COMBUSTION HAZARD

FIGURE 4

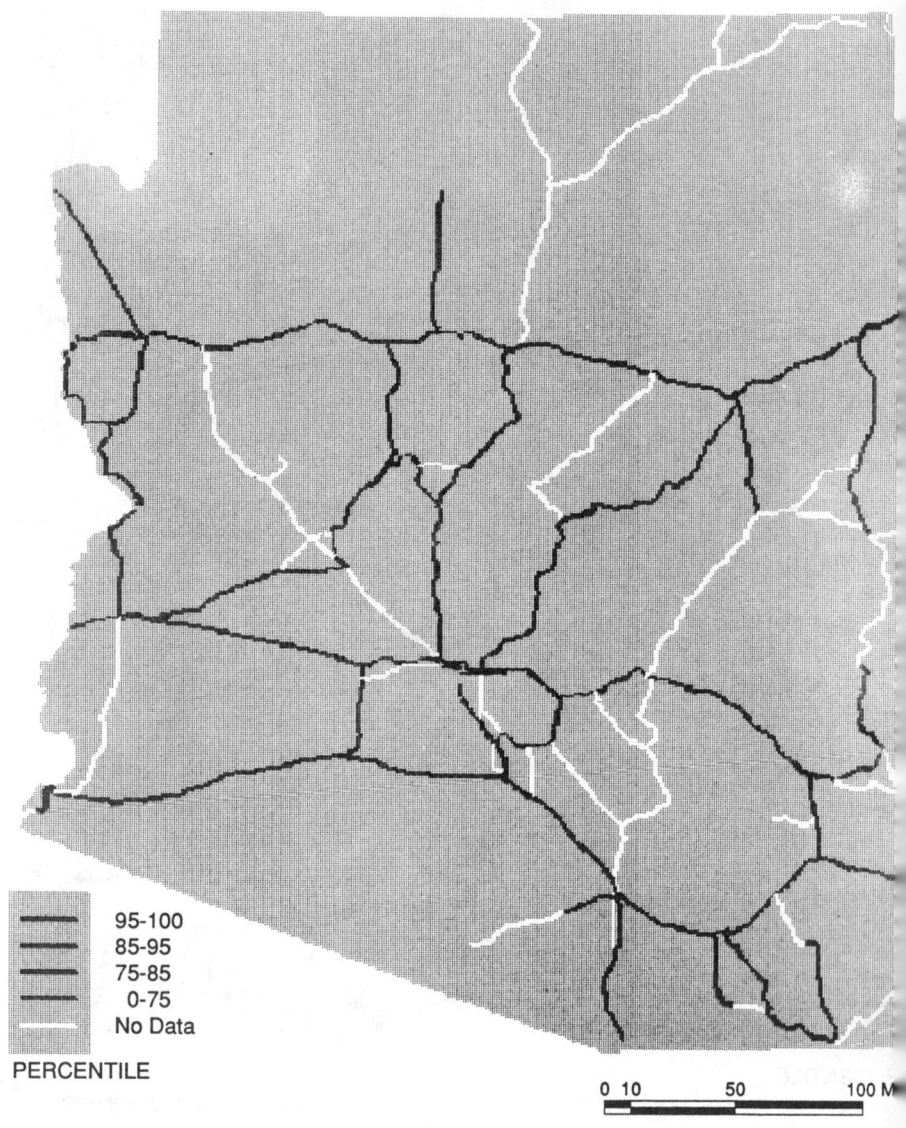

COMBUSTION RISK
FIGURE 5

parameters a maximum of 76 minutes for any segment of the network was obtained and a maximum of 136 minutes for any point in the state.

Figure 6 presents the results of the response time analysis. With this information, the model can calculate the relative vulnerability for each route segment by multiplying the truck accident rate by the volume of hazardous material by the population affected by the response time index. A response time index which ranged from 1 to 5 was used instead of the actual response time in minutes because using actual response time gave too much weight to response time vis a vis population at risk. The response time index was calculated as follows:

Response Time Index	Actual Response Time
1	Less than 15 minutes
2	10 to 30 minutes
3	31 to 45 minutes
4	46 to 60 minutes
5	> 60 minutes

The results from analysis of different types of hazardous materials can be combined to obtain a combined or composite risk. The model has the ability to weight the outputs of the analysis of different types of hazardous materials depending on their relative hazard. Other variables, such as the sensitivity of the vulnerability index, can also be easily changed to calibrate the model to specific situations.

Application of GIS Analysis

The potential applications of the GIS analysis is very broad. Because of its flexibility and the extensive special databases that are being developed, it will likely be the primary risk analysis tool of the future.

Public sector applications could include:

o Highway construction and maintenance prioritization
o Routing of hazardous materials and waste
o Transportation mode alternative analysis
o Siting emergency response units
o Assessing and prioritizing training for emergency response units

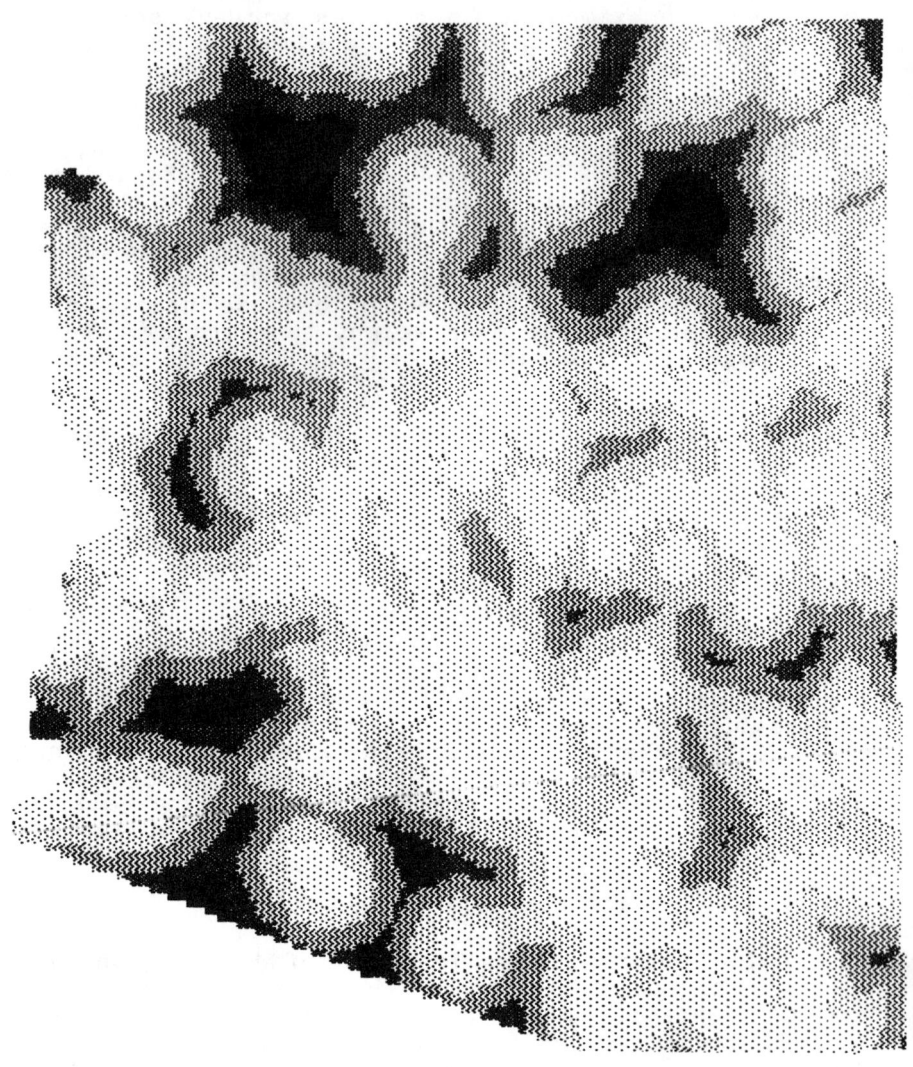

EMERGENCY RESPONSE TIME CONTOURS

FIGURE 6

- o Evaluating risks to sensitive population centers
- o Evaluating risks to sensitive ecological areas

There are also a number of private sector applications of the GIS risk analysis model:

- o Minimum time and mileage routing
- o Improved utilization of equipment and personnel
- o Minimum population exposed routes
- o Time of day risk analysis routing alternatives

These types of analysis can be done on a national, state, regional or local basis. The GIS also has the capability to "zoom in" and enlarge a specific area of interest, thereby allowing a more detailed evaluation of areas of particular interest.

Conclusions

The GIS risk analysis proved very successful in its application to the Arizona highway system. In addition to a relative risk evaluation for each type of hazardous material, the system was able to assess the vulnerability of each highway segment bases of the proximity of the emergency response units and the response time.

One of the criticisms of traditional link-segment model has been that the segments are too long and the demographic data are not detailed enough. With increased computing power and more detailed GIS databases becoming available, the GIS will allow for a much more detailed analysis than the traditional methods. These factors, in addition to the obvious graphic display benefits, will likely result in the GIS becoming the primary risk analysis tool of the future.

Use of Advanced Technologies for Improving Hazmat Transportation Safety

Dr. Ashok B. Boghani[1]

Abstract

This paper describes how advanced technologies can help improve the safety of over the road hazardous material transportation. The technologies discussed include driver information systems, vehicle control, surveillance and communication systems. Several specific ideas on how to evaluate the benefits of these technologies are also presented.

1. Introduction

The transportation industry is undergoing major change. The "Information Age" promises to have a dramatic impact on the way the transportation function is being performed today and likely to be performed in the future. Over the past few years, a variety of new products and services have become available, and more are being developed each year. These products and services reduce the capital and manpower requirements of the transportation infrastructure, and improve the quality and safety of transportation.

These developments are particularly significant to the transportation of hazardous materials (hazmat). As a society, we are becoming increasingly aware of the risks involved in hazmat transportation. We want to protect lives and property, as well as our environment, and as we exercise tighter control on the movement of hazmat, we can use all the help we can get. Advanced technologies provide such help.

However, we need to recognize that advanced technologies are not the answer to all of the problems, because transportation safety can be enhanced through operational improvements as well. For example, improved routing, greater control over the allowable time of travel, and better training can also reduce risk significantly. In these instances, technologies play more a complementary role.

This paper will concentrate on areas where technological advances do have a significant impact. Some of these areas are communication, driver information systems, vehicle control, and surveillance. (There are numerous references dealing with these technologies, e.g., Grenzeback et. al. 1988, Boghani 1989, USDOT 1990, Willis 1990, Schneiderman 1989, and reports published by Mobility 2000.) The technologies that perform these functions are generally related to the collection, processing, and transmission of information. In this paper we will discuss some of these technologies, their contribution and improved safety, and the ways in which their impact can be evaluated. This discussion is restricted to truck transportation of hazmat, but other modes can equally benefit from these technologies.

---

[1]Director, Logistics and Transportation Management Unit, Arthur D. Little, Inc. , Cambridge MA.

2. Applicable Technologies

The risks involved in the transportation of hazardous materials arises primarily from release of such materials caused by accidents. A whole chain of events takes place before people are exposed to the harmful effects of hazmat—i.e., the accident takes place, material is released, it spreads (or explodes) and people are caught in the way. To combat this risk, we can attack the chain at various places. We can:

- Reduce the likelihood of the accident
- Reduce the likelihood of a release in a given accident
- Reduce the likelihood of severe consequences given a release, or
- Reduce the likely impact on people given a consequence

The advanced technologies assist in each of these strategies in a varying degree.

Reducing the Likelihood of an Accident

There are various ways in which advanced technologies can help reduce the likelihood of an accident. The paragraphs below describe several technologies dealing with vehicle control, driver information system, driver/vehicle performance measurement and heavy truck detection.

Vehicle Control. The technologies in this area assist the driver in controlling the vehicle. One of the more promising developments is a collision avoidance system. This involves tracking a vehicle in front using a radar or a laser. The distance between the two vehicles and their relative speeds are measured. If an imminent collision is expected, the system can alert the driver or apply the brakes automatically. Although these types of devices have been available for some time, they have not been perfected to the point that both false positive or false negative indications are completely eliminated. Also, there is the issue of liability—can the manufacturer face a product liability suit if the driver relies on such a system and gets into an accident?

The application of the device can be extended so that the vehicle is able to maintain a constant distance between itself and the vehicle in front through judicious use of the brakes and the cruise control system. When perfected, this would do wonders to reduce driver fatigue, which is a cause of accidents.

Driver Information Systems. A variety of technologies are being developed to improve the quality of information being provided to the driver. One major group of technologies in this area relates to informing the driver where he is in relationship to his objective. This requires finding the vehicle's location (in an absolute term), and superimposing this location on an electronic map to locate the vehicle with respect to the destination.

There are five technologies that fit into the category of automatic vehicle location (AVL).

1) *Dead-reckoning system*—the orientation and distance traveled are monitored, and, given an initial position, the system can find the vehicle's current location. An example of such a system is one made by ETAK.

2) *Ground-based radio-determination system*—an example of this type is LORAN-C, which has been available for navigating in coastal waters for quite a few years, and is now available for navigation over land. This system is used extensively in services offered by firms such as Qualcomm, Geostar and Motorola, and employs the principle of triangulation, as shown in Figure 1. Users seem generally satisfied with this system, but in certain areas, the radio signal can be weak, making location somewhat uncertain.

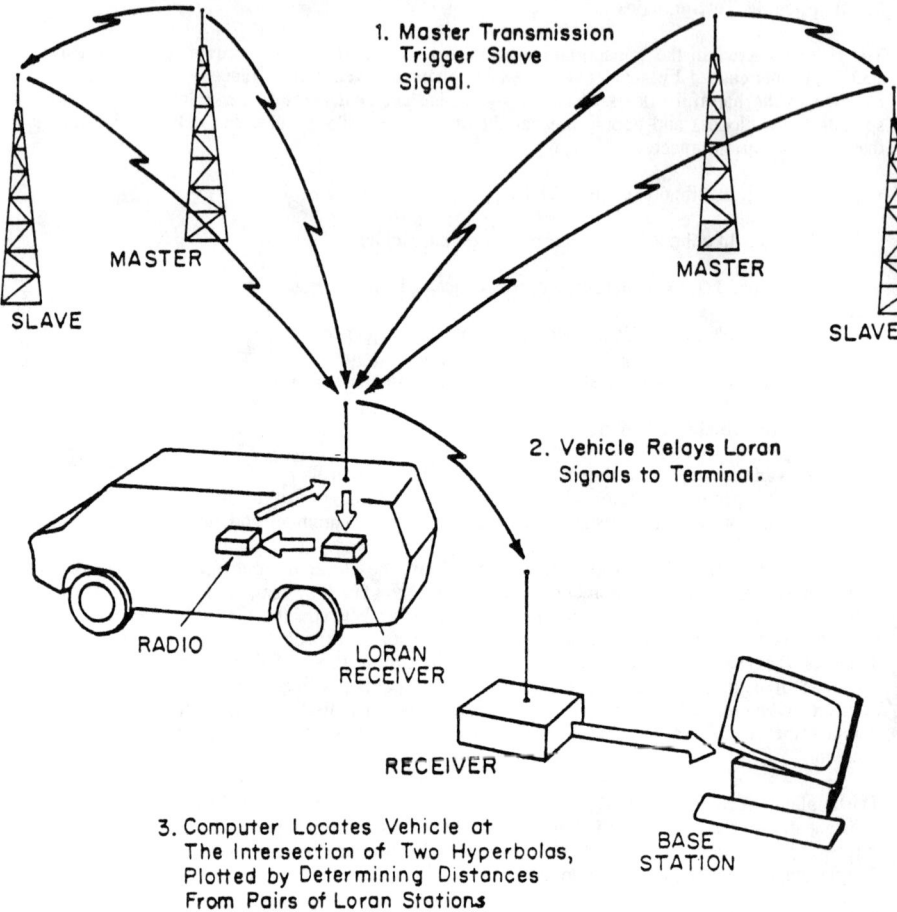

Figure 1: Loran-C Vehicle Location System

3) *Low-earth orbit satellite-based system*—the only candidate in this category is Global Positioning Systems (GPS/Navstar), developed by the Department of Defense. This system also employs triangulation, but the points from which distances are measured are satellites, and not radio-masts as used by LORAN-C. The required calculations are performed on a vehicle-based device that is fairly expensive. In addition, not all the required satellites are in orbit yet so there can be service gaps, depending on the time of the day that this service is available.

4) *Radio determination satellite systems (RDSS)*—these systems use distances from two satellites in geosynchronous orbit to determine location. The calculations are performed at a central facility and the results are transmitted back to the truck (Figure 2). Two firms, Geostar and Qualcomm, are planning to determine location based on this principle eventually. However, they both currently use LORAN-C to perform this task.

5) *Proximity system*—in this system, fixed devices en route are used to determine the location of trucks carrying transponders (often called radio tags). The readers can be mounted on signposts, traffic lights or on the ground. The obvious problem with such a system is that the location is known only at the points where readers are located. Unless a dense network of readers is deployed, the truck drivers will not be able to rely on such a system for determining location.

To complement AVL technologies, databases are now available to display city maps on a video screen. These electronic maps are generally provided in the form of a compact disc (much like the ones used to record music). The present location of the truck can be displayed on such an electronic map to assist the driver navigate and therefore drive more safely.

The capabilities of a driver information system of the type described above will get a major boost when real time information on the status of infrastructure becomes available. Some day in the future, there will be information on traffic congestion, road repair, speed restrictions, slippery conditions and soon will be made available on a real time basis (Agnew 1988). The drivers will then be able to make informed decisions regarding the safety of truck operation and avoid getting into trouble.

In terms of displaying information in a better manner, experiments are underway to equip cars (and trucks) with heads-up-displays (HUD) working much like their counterparts in fighter aircraft. These displays project necessary information in front of the driver so that he does not have to take his eyes away from the road to read the instrument panel.

Driver/Vehicle Performance. The trucks can now be equipped with on-board computer/vehicle management system (OBC/VMS) to monitor the performance of trucks and drivers. Parameters such as hours on the road, vehicle speed, and time since last break can be measured and recorded. The vehicle maintenance parameters can also be measured and recorded if so desired. This information can be conveyed in real time to a dispatch center in order to monitor the performance of the driver/vehicle continuously. The result—a safer operation.

Heavy Truck Detection. One potential cause of unsafe operation is truck carrying more load than it should. Such overweight trucks can be detected using a weigh-in-motion (WIM) system. This system is installed in the pavement, as shown in Figure 3, and it provides measurement of weight of each vehicle without requiring it to slow down. In most applications, a WIM system is installed together with an automatic vehicle classification (AVC) system. The two systems working together can not only weigh a truck but classify it and determine if the truck can operate safety on a bridge (i.e., comply with the "bridge formula").

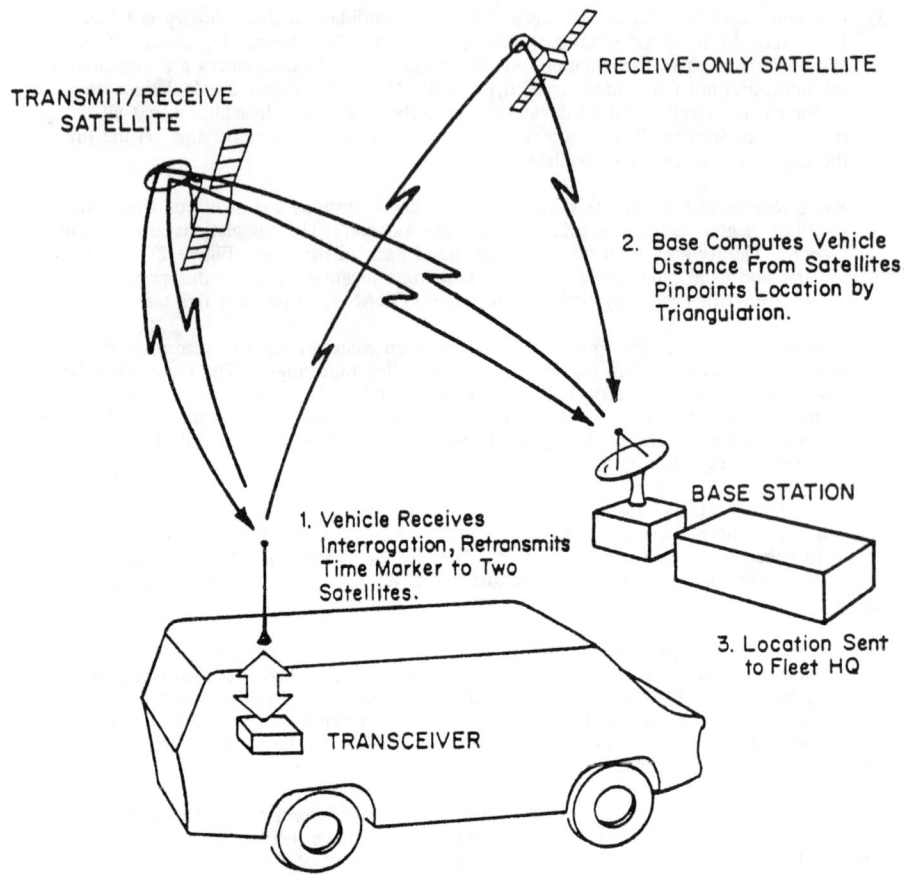

Figure 2: Geostar Radio-Determination System

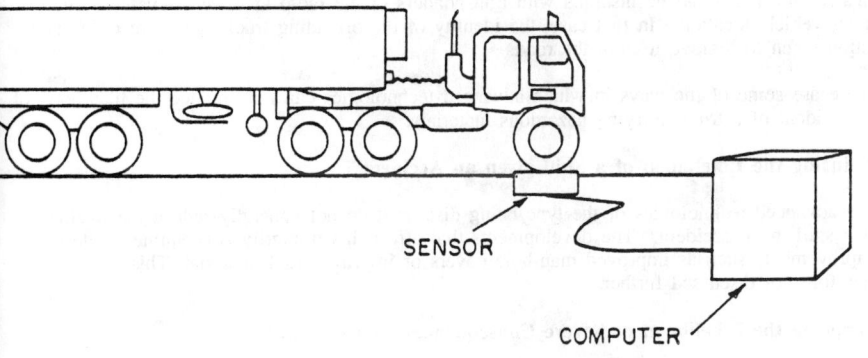

Figure 3: Schematic Diagram of a Weigh-In-Motion System

Finally, the trucks can be installed with transponders or the radio tags we discussed earlier under vehicle location. In that case, the identity of the offending truck can be recorded and action taken to remove it from the road.

These are some of the ways in which advanced technologies can help reduce the likelihood of an accident of a truck carrying hazardous materials.

Reducing the Likelihood of a Spill given an Accident

The advanced technologies of the type being discussed do not generally reduce the likelihood of a spill in an accident. The developments that affect this primarily concentrate on design improvements such as improved man-hole covers or improved tank material. This issue is therefore not discussed further.

Reducing the Likelihood of Severe Consequences Given a Spill

The consequences of some types of hazmat spills can be kept to a minimum if immediate and correct actions are taken. For this to happen, the driver may need to communicate immediately with his dispatcher or an emergency response center. Technologies that provide instant two-way communication can help under these circumstances.

Some of the products mentioned earlier provide two-way communication service in addition to helping locate the vehicle. For example, both Geostar and Qualcomm allow a two-way exchange of alphanumeric messages. In an emergency, a message can be transmitted simply by pressing an emergency key. The dispatcher would not only know that the driver needs help, but also the location of the truck. Help can be sent right away and the driver can also be informed of the best way to respond to the spill in order to prevent more severe consequences.

Reducing the Likely Impact on People

Given that a significant release has taken place, can the advanced technologies reduce the impact on people? The answer is yes. Once again, communication is the key. The information on the spill of a toxic material and the need to evacuate people can be transferred rapidly from the truck to an emergency response center to the local police or fire department. A prompt evacuation can then be ordered before harmful vapors reach the population.

Often, the victims of such spills are not people living or working nearby, but those who happen to be driving by in automobiles. If the infrastructure exists to transfer information to the drivers using a driver information system of the type mentioned earlier, these motorists can also be protected from the impact of spills.

Advanced technologies can help improve safety of hazardous material transportation in a variety of ways. Now let us examine how the benefits of these measures can be evaluated.

3. Evaluating the Benefits

In order to make educated decisions on whether advanced technologies should be deployed or not, it helps to first quantify the potential benefits. This statement needs several caveats. First, most of these technologies provide other benefits besides improving safety. For example, two-way communication and location services help improve the productivity of operation as well as safety, so the safety benefits are only a part of the overall benefits. Second, the decision may not be to put a value on safety, but rather to take all possible measures to reduce the risk to the greatest extent possible. In this case, there is no need to evaluate safety benefits. These two issues notwithstanding, here is a brief discussion on evaluation of benefits.

The most promising method to evaluate benefits is risk assessment. This fairly widely used technique can be applied to quantify risk before and after the application of a particular technology. The first step in performing such an analysis is to identify whether the technology affects the *frequency* of release or the *consequences*. Needless to say, some technologies that affect the likekihood of an accident change release frequency, while others affect the consequences.

In either case, the process of quantifying the impact is not easy. For example, what will be the percentage reduction in accident frequency if the driver has accurate location information available to him? Traditionally, the risk analysis methodology has relied on two sources for such estimation: 1) past history, and 2) engineering analysis. In order to use historical results, there has to be historical data, which is not the case for technologies that have yet to see a widespread use. Engineering analysis is the second choice. For example, for the case cited above, an estimation on likely frequency reduction could be made based on human factors research, coupled with lab tests on simulators. Similarly, the likely benefits of a faster emergency response can also be estimated, based on factors such as how fast a toxic vapor cloud would spread and how the number of people in a given area could be reduced under different emergency response strategies. Using this methodology, the benefits can be quantified and appropriate decisions on technology deployment can be made.

4. Conclusion

This paper has briefly described some of the advanced technologies that can be used to reduce risks involved in transportation of hazardous materials. As the use of these technologies become more and more widespread, society will start feeling their impact. This is not likely to take too long now—by the time this decade comes to a close, every technology discussed in this paper should be in routine use.

References

Agnew, W.G., "Future Personal Ground Transportation," GMR-6419, GM Research Laboratories, Warren, MI, September 1988.

Anon., "Advanced Driver Information Systems," Mobility 2000, March 1990.

Anon., "Commercial Operations," Mobility 2000, March 1990.

Anon., "Advanced Vehicle Control Systems," Mobility 2000, March 1990.

Anon., "Report to Congress on Intelligent Vehicle-Highway Systems," U.S. DOT, Office of the Secretary of Transportation, DOT-P-37-90-1, March 1990.

Boghani, A.B., "Automation Thrusts in Rail and Road Transporatation," Fifth World Conference on Transportation Research, Yokohama, Japan, July 1989.

Grenzeback, L.R., Stowers, J.R., and Boghani, A.B., "Feasibility of a National Heavy-Vehicle Monitoring System," NCHRP Report 303, Transportation Research Board, PB89-182661/WTS, Washington, DC., December 1988.

Schneiderman, Ron, "Tracking Trucks by Satellite," High Technology Business, May 1989.

Willis, D.K., "IVHS Technologies: Promising Palliatives or Popular Poppycock?," Transportation Quarterly, Vol. 44, No. 1, Eno Foundation, Westport, CT, January 1990.

StateGEN/StateNET and DOT Guidelines:
Tools for Highway Routing of Hazardous Materials

J. W. Cashwell[1], J. D. Brogan[2], C. M. Erickson[1]

Abstract

Under the sponsorship of the U.S. Department of Energy Office of Defense Programs, Sandia National Laboratories' Transportation Technology Center has developed computerized software to assist state and local governments in determining highway routing alternatives for radioactive materials. The techniques have a still wider application in analyzing route alternatives for all hazardous materials transport activities. The specific techniques discussed in this presentation are the latest update of the StateGEN/StateNET model structure and routing algorithm which contains the codified U.S. Department of Transportation (DOT) Guidelines for Highway Route Controlled Quantity shipments of radioactive materials.

Introduction

StateGEN/StateNET, a modelling structure and routing algorithm designed expressly to assist state and local governments in analyses of routing alternatives, was released in 1989. StateGEN/StateNET was designed to permit the user to construct a network and assign attributes of interest to links in the network on a personal computer (PC). The completed network is then transferred via a modem to the TRANSNET system (Ca89), and a preferred route is determined based upon attribute weights assigned by the user.

This modelling structure permits a state or local government to perform an analysis of routing alternatives based on weighted attributes that reflect local concerns and available data. The StateGEN/StateNET software has recently been enhanced to incorporate the ability to

---
1. Sandia National Laboratories, Albuquerque, New Mexico
2. University of New Mexico under contract to Sandia National Laboratories

perform a routing analysis consistent with that recommended by the U.S. Department of Transportation (DOT) for Highway Route Controlled Quantity shipments of radioactive materials (US89) with a minimum of resources.

Sandia National Laboratories (SNL) is the Department of Energy's (DOE) lead organization for transportation research and development. The DOE Office of Defense Programs has been the prime sponsor of development of models and associated databases used to analyze the impacts of the transportation of radioactive materials. The StateGEN/ StateNET software was based on the existing models on TRANSNET, a system that was developed to give outside users access to analytical codes and associated data developed for the DOE.

Background of the Transnet System

TRANSNET is being used to support DOE site environmental analysis, risk assessments and systems analyses for the defense and repository programs, routing assessments for the DOE and states, and operational analyses as well as basic research.

The goals of the TRANSNET system are to speed transfer of transportation risk and systems analysis technology to the public sector by permitting users to access the most comprehensive and up-to-date transportation risk and systems analysis models and associated databases. Models are updated and refined as information and techniques become available. TRANSNET provides the user with the latest version of these models and databases in a timely manner. Users of the TRANSNET system are allowed to construct input files and to use or modify input files previously developed for DOE-sponsored analyses. User operating and equipment costs are minimized by establishing the TRANSNET system on a centralized computer and allowing access via a modem-equipped personal computer. This realizes another prime goal, which is to develop and operate the TRANSNET system with a maximum of flexibility while minimizing system costs.

Prior to placement on the TRANSNET system, codes are modified to incorporate a user-friendly interface, if one did not already exist. Input data from analyses performed by SNL for the DOE are structured to allow their use either directly or with editing. The user is also permitted to construct an input data set. Output files from each of the codes are structured to permit variable levels of detail that correspond with user requirements. In addition, interfaces between the codes and data sets are built to permit direct data transfer between codes.

General Background

The DOT has issued two guidelines documents to address transportation of hazardous materials. 49 CFR Part 397 states that "Unless there is no practicable alternative, a motor vehicle which contains hazardous materials must be operated over routes which do not go through or near heavily populated areas, places where crowds are assembled, tunnels, narrow streets, or alleys. Operating convenience is not a basis for determining whether it is practicable to operate a motor vehicle in accordance with this paragraph." To assist in determination of eligible routes, the DOT has issued two documents, one entitled "Development of Criteria to Designate Routes for Transporting Hazardous Materials--Final Report" (US80a) and an accompanying guide, entitled "Guidelines for Applying Criteria to Designate Routes for Transporting Hazardous Materials" (US80b). The DOT further specifies in 49 CFR Part 177 that Highway Route-Controlled Quantity shipments of radioactive materials should travel from origin to destination via the shortest Interstate highway route unless alternative route(s) have been designated by the state(s) through which the shipment would pass. The DOT issued a document entitled "Guidelines for Selecting Preferred Highway Routes for Highway Route Controlled Quantity Shipments of Radioactive Materials," which provides states with a methodology for assessing potential alternative routes through the assignment of primary and secondary factors. The latter are determined locally, but the methodology employed by the state routing agency to determine an alternative route must be consistent with that recommended by the DOT. Each state interested in determining alternative route(s) must interact with adjoining states to assure continuity of the transportation network.

The primary goal of the StateGEN/StateNET system is to assist state and local governments in this effort by providing a structured technique consistent with models used in national analyses. Through the use of the TRANSNET system, state or local governments can benefit from the analytical capabilities of the models, including indicators of the radiological and nonradiological risks of transport. This is accomplished through a two-part methodology; the first, entitled StateGEN, allows the user to generate a transportation network of interest, construct the network on a personal computer, and assign specific attributes to the network. The network is defined by nodes (uniquely numbered geographic locations such as intersections or boundary lines), with links (pathways) connecting the nodes. The particular attributes of interest are determined by the user; however, it is assumed that the DOT route selection methodology will both be of interest and be used to assist in the determination of particular attributes. After gathering the necessary data, the user assigns attribute values to individual links consistent with the structure provided in the StateGEN system. Link-specific attributes could include direct data for the link as well as calculated attributes, such as the risk parameters mentioned above. The DOT methodology categorizes these into "primary and secondary factors" which correspond respectively to objective attributes and subjective attributes of interest to a specific jurisdiction. These network data are structured on a personal computer in a form consistent with the StateNET model to minimize TRANSNET hookup time and costs. StateGEN also permits the user to check the network for continuity and to perform single-attribute route selections. StateGEN is distributed to the user on a flexible 5 1/4" floppy diskette.

After development of a structured network and associated attributes, the user transfers the files to StateNET on the TRANSNET system. StateNET allows a user to maximize or minimize each network attribute of interest and to assign a weight to each. The StateNET model may then be used to determine the route that best represents the user-assigned weights for all characteristics. In addition, StateGEN can be used to structure input data files for the RADTRAN risk assessment model, which will calculate link-specific risks for the user's network; the StateGEN network can then be edited to include these factors. This allows the user to calculate route alternatives on StateNET using risk parameters specific to the network.

Discussion

StateGEN/StateNET Objectives

The StateGEN/StateNET code and User's Guide were designed to meet the following needs:

* To build models of transportation networks.
    The networks may be used to calculate and compare all possible routes through a region. The region size and network are determined by the user. Networks are described by identifying nodes and links. The length of each link (i.e. the distance between two nodes) is required input.

* To consider attributes of the transportation network.
    It is assumed that factors of local concern other than distance will be used to determine route alternatives. Attributes regarding the characteristics of each of the links on the network are assigned by the user. These attributes could be data and/or calculated parameters from either the models on TRANSNET or other sources.

* To fit available data accurately to the transportation network.
    The task most challenging to the user is how to extract a specific value for each link from the available data, which may not have been collected in a form that is directly compatible with the design of the network.

* To make temporary changes to a network model.
    In the analysis of a single problem, different configurations of the same network may be examined in order to calculate different sets of optimal routes. StateNET allows weighted optimization of properties to be combined with temporary reconfigurations of the network (i.e., temporary inclusion/ exclusion of certain links), creating a powerful and flexible tool for routing analysis.

* To maintain a current network.
    Over time, network models require updating and modification. The network must accurately describe the current status of the link attributes to be of use in future analyses. StateGEN allows for rapid retrieval and editing of all network data.

* To assist in applications of the DOT Guidelines
    A separate software package that was developed to assist states and local governments in

application of the DOT Guidelines for Selecting Preferred Highway Routes for Highway Route Controlled Quantity Shipments of Radioactive Materials is available to StateGEN users. It provides a structured method of compiling the necessary data and reproducing the intermediate calculations and required worksheets on the user's personal computer.

The StateGEN Module

StateGEN is structured for use on a IBM-compatible personal computer. The StateGEN diskette and user guide are mailed upon request, free of charge, to TRANSNET subscribers. StateGEN allows the user to create data files needed to construct a model of a transportation network. The files needed are:

a. a file of nodes to identify the geographic boundaries or "break points" on the network (these are most commonly described as intersections, particularly for the roadway system);

b. a file of links to identify the network connections between nodes;

c. a data dictionary file to list the attributes of links that are of interest to the user, e.g., the primary and secondary factors described in the DOT methodology.

The primary StateGEN program, FILES, uses menus and windows to add, edit, and delete data in these files. Data entry and network maintenance are minimal. The same data item is never entered twice. Validation of the network model may be performed at any time while FILES is in use. Modifications needed when a network is changed are performed automatically. For example, when the user adds an attribute to the data dictionary, a file that reflects the current network is automatically created to hold the data for that attribute for each link on the network. When the user changes the network, the attribute file (along with other attribute files) is automatically updated to reflect the new network.

The user should:

a. determine what attributes are of interest,
b. collect data about these attributes,
c. use these data to assign a specific numeric value to each link for each of the defined attributes.

The programs provide assurance to the user that the network model contains reasonable data and is contiguous. The validation option performs internal checks on the data fields required for proper operation of the routing algorithm. If a set of data files passes inspection, it represents a valid network and is valid to use as input to StateNET.

The StateNET Module

StateNET is a routing program that resides on TRANSNET. StateNET input data are the files developed using the StateGEN program. These files are uploaded from the user's personal computer to TRANSNET, where they reside in the user's personal directory. No other TRANSNET user has access to them. Networks may, however, be shared with other TRANSNET users by special arrangement with SNL.

StateNET permits modification of a network. To change the optimal route without changing the weighting of the attributes, links may be temporarily ignored by changing their routing status from active to inactive. An inactive link is unavailable to the routing algorithm.

To indicate which attributes should dominate the analysis, the user assigns relative weights to each parameter of interest. This ability to weight attributes greatly increases the number of possible outcomes from the routing algorithm. When a route is calculated by optimizing (minimizing or maximizing) only one parameter, then the raw number assigned to each link can be used as the optimizing value. The minimum-value route is then determined by the StateNET algorithm.

When more than one parameter is used to calculate a route, the parameters are normalized prior to assigning weights. The normalized values are multiplied by the user-assigned weights for the parameters and divided by the sum of the weights given to all parameters. A single frequency distribution over all parameters is then used by the StateNET routing algorithm to determine the route.

The DOT Radioactive Materials Routing Guidelines

To more directly assist states and local governments in their efforts to comply with the DOT guidelines that apply to Highway Route Controlled Quantity shipments of radioactive materials, Sandia has created a computerized version of the methodology. The user defines the network alternatives to be analyzed and provides the route-specific data required by the methodology on a spreadsheet. The risk and cost equations described in the guidelines are directly used to reproduce the worksheets. The user has the option of printing all the worksheets described, or the summary sheets. For clarity, additional forms were created for some of the intermediate calculations.

The DOT guidelines document recommends that the described methodology does not represent the only method of conducting an adequate routing analysis. States are permitted to use "an equivalent routing analysis which adequately considers overall risk to the public." To facilitate the development of an equivalent method, users of StateGEN/StateNET and the TRANSNET system can apply other models and databases available on the system. For example, RADTRAN, one of the codes available on TRANSNET, is used by the DOE to estimate radiological and nonradiological risks of transport in support of environmental analyses. If sufficient network data are available to support the use of RADTRAN, the TRANSNET user can perform a comparative routing analysis using risk estimates directly from the code.

Development/Applications

StateGEN/StateNET was developed with input from the State of Nevada Agency for Nuclear Projects, Nuclear Waste Project Office and their contractor, University of Nevada, Reno. SNL then worked with the State and its contractor to tailor and refine the software prior to its general release. Nevada subsequently applied the StateGEN/StateNET software to assist in the designation of candidate route alternatives for further analysis (Co89). The State has announced that it will base its further analysis on the revised StateGEN/StateNET software incorporating the DOT Guidelines.

A draft copy of the revised software incorporating the coded DOT Guidelines discussed above was provided to the State of New Mexico for use in the designation of routes for shipments to the Waste Isolation Pilot Plant, a DOE facility designated for storage of transuranic and some high level nuclear wastes generated as byproducts of weapons-production activities. The State's contractor, the University of New Mexico, used the coded guidelines to perform an analysis of over 40 route alternatives (Br90). The University also provided input to SNL regarding application of the package and development of user-friendly interfaces.

Following incorporation of these comments and recommendations, StateGEN/StateNET will be reissued. This release is currently scheduled for July, 1990.

Conclusions

StateGEN/StateNET2 with DOT Guidelines allow local governments and other interested parties to perform a detailed transportation network analysis using methods developed for the DOE's defense waste management program. Together with other capabilities available on the TRANSNET system, the user can perform network optimization consistent with the requirements of the DOT as well as local-area interests.

Equipment requirements are consistent with those of TRANSNET: an IBM-compatible personal computer, Hayes-compatible modem, telephone, and communications software.

Parties interested in using StateGEN/StateNET on the TRANSNET system should submit a written request for access, including contact person, sponsor and intended use to:

> J. W. Cashwell
> Division 6321
> Sandia National Laboratories
> P.O. Box 5800
> Albuquerque, New Mexico 87185

References

Br90   J.D. Brogan, Statement for the Public Record Regarding Proposed Designation of New Mexico Highway Routes for Shipments of Transuranic Radioactive Wastes to the WIPP, University of New Mexico, Albuquerque, NM, April 9-18, 1990.

Ca89   J.W. Cashwell, TRANSNET--Access to Transportation Models and Databases, SAND89-0982C, TTC-0882, Sandia National Laboratories, Albuquerque, N.M, June, 1989.

Co89   M.V. Ardila-Coulson, The Statewide Radioactive Materials Transportation Plan, Phase II, University of Nevada, Reno, December, 1989.

US80a U.S. Department of Transportation, Development of Criteria to Designate Routes for Transporting Hazardous Materials--Final Report, DOT/FHWA/RD-80/105, July, 1980.

US80b   U.S. Department of Transportation, Guidelines for Applying Criteria to Designate Routes for Transporting Hazardous Materials, DOT/FHWA/IP/80/15, November, 1980.

US89   U.S. Department of Transportation, Guidelines for Selecting Preferred Highway Routes for Highway Route Controlled Quantity Shipments of Radioactive Materials, DOT/RSPA/OHMT-89/01, January, 1989.

COMPUTER-ASSISTED RISK ASSESSMENT OF DANGEROUS GOODS TRANSPORTATION FOR HAUTE-NORMANDIE

Sylvain LASSARRE [2], Kurt FEDRA [1], Elisabeth WEIGKRICHT [3]

Abstract

To assess the "best" routes for the transport of dangerous goods in an area of 600 km2 in France, software based on a geographical information system to manage, treat and represent statiscal and geographical data related to the evaluation of the risk of transport on a road network was developed.

Probability of accident and release are attributed to arcs of the digitized network provided by IGN [4] on a 1/50 000 scale, according to their characteritics. Active, scholar, and non-active population issued from the 1982 INSEE [5] census, are distributed among categorized (urban, industrial...) areas of administrative units called "communes". The intersection of this population layout with an impact zone (a corridor along the road of which the width depends on the nature and quantity of the dangerous materials), leads to a measure of the exposed people to damage.

Multiplication of the accident probability by exposed people gives an estimation of the relative risk used to compare alternative routes in order to choose the less risky route.

---

[1] Head, Advanced Computer Applications, International Institute for Applied Systems Analysis, A-2361, Laxenburg, Austria.

[2] Researcher, Département d'Evaluation et de Recherche en Accidentologie, Institut National de Recherche sur les Transports et leur Sécurité, BP 34, 94114 Arcueil Cédex, France.

[3] Researcher, Advanced Computer Applications, International Institute for Applied Systems Analysis, A-2361, Laxenburg, Austria.

[4] Institut Géographique National (France).

[5] Institut National de la Statistique et des Etudes Economiques (France).

I - Background and problem addressed

The last OECD group on the Transportation of dangerous goods (OECD, 1988) emphasized the interest of computer-assisted management of the risk generated by the transport of dangerous goods and especially its application to routing.

INRETS, funded by the French Mission du Transport des Matières Dangereuses from the Direction des Transports Terrestres, has set a research objective to develop software based on a geographical information system for the assessment of the risk of a road network in collaboration with the research team of the Advanced Computer Applications of the International Institute for Applied Systems Analysis. Current application areas dealt with by the group include technological risk assessment, transportation risk/cost analysis, and a decision-oriented software package with a graphics interface, for the management of hazardous subtances and industrial risk, was completed in 1988 (IIASA, 1988).

As a feasibility study, this prototype should be applied as a decision-aid tool to look for safest route in an area situated in Haute-Normandie (Fedra et al., 1989). In this area, which covers 43 communal districts, there are sixteen chemical and oil-industry firms which generate an important flow of dangerous goods by road, rail and water.

For the time being, there are no special state regulations concerning the routing of dangerous goods in France. Only at the local level, "the prefets" and the mayors, by means of the police authority, control for the movement and parking of vehicles transporting hazardous materials.

Traffic in tunnels, and on motorways around Paris are strictly regulated by law. Some limitation measures are taken at a state level to forbid the traffic of dangerous goods at certain times of the year, especially during weekends and holidays.

The first studies concerning the assessment of risk of transportation of dangerous goods apart from nuclear materials were undertaken a few years ago in the vicinity of Lyon for some products (Hubert and Pages, 1986, 1989).

II - An approach to risk assessment of dangerous goods transportation

In general, risk evaluation should involve an assessment of the following :

- the potential hazards of the substances involved ;

- the packaging aspects ,

- the vehicle and route selection in relation to accident probabilities (eg. road category, level crossings, bridges and tunnels, etc.) ;

- the operational aspects of transportation, such as speed limits, police escorts, restrictions to daytime mobilization, or certain weather conditions, etc. ;

- the environment around the network in terms of population density, and in particular, vulnerable installations such as schools or hospitals, general land use, and in particular, sensitive, environmental zones, such as water resource areas.

II.1 - Schematic representation of probabilistic risk analysis

Based on the schematic probabilistic risk analysis (OECD, 1988), the analysis of transportation flows, of accident scenarios and effects are related to the network, whereas the exposure analysis is related to the surroundings of the network.

The risk is generated by a chain formed by a dangerous substance transported by a vehicle along a network through a territory. The vehicles transporting dangerous substances induce flows measured in number of vehicles or transported tons. The network is cut into sections to which an accident rate is attributed according to its characteristics. Depending on the nature of the substance, an impact zone round the route contains the population exposed to risk on the territory. The product of the accident rate by the number of persons exposed along the route in the impact zone, gives the value of the risk which is taken as the criteria to be minimized in order to search for the safest route (Pijawka, 1985) (Saccamano, 1985) (Kessler, 1987). The variability in spill probabilities for different accident conditions which generates a risk profile instead of a single measure, is not taken into account.

In order to evaluate feasible alternatives, and find acceptable compromises between risk and economic criteria, the simulation of activities related to transportation requires not only detailed information about the transportation network in question, but also geographical and environmental data. These categories lend themselves to representation and manipulation by means of Geographical Information Systems (GIS), coupled with relational data base structures for the non-spatial data categories.

II.2 - Modeling accident rates and consequences

For an arc of a network, the accident rate is equal to the frequency of occurrence of an accident involving a vehicle transporting hazardous materials, multiplied by the length of the arc. This rate λ is usually expressed as a number of accidents per vehicle times kilometers, and depends on the road category.

The direct estimation of the rate λ is difficult because of the lack of statistics for accidents involving hazardous materials, and the lack of information about the number of vehicles transporting hazardous materials. To estimate this rate λ, the heavy vehicle accident rate for different road categories and locations (rural and urban areas) is evaluated and multiplied by a factor specific to dangerous goods, usually less than 1, to obtain the rate of accidents involving heavy vehicles transporting dangerous goods (Hubert, Pagès and Degrange, 1986). This leads to the estimates of accident rates on four road categories : motorways, national roads, secondary roads, others.

motorways	2.5
national roads	7.5
secondary roads	12.5
others	12.5

Table 1 : Rate of accident (10^{-7}) involving hazardous materials by road categories

The consequences C are defined by the integral, over an impact zone E, of the density of the exposed object, in our case the population : $C = \int_E \rho(e)\, de$.

The exposed object could be :
- concentrated on a site,
- distributed along an arc,
- distributed over an area.

If the density is constant over the impact zone, then $C = \rho \times \mu(E)$ with μ being a measurement of E ;
- if E is a site, then $\mu(E) = 1$,
- if E is an arc, then $\mu(E) = L$, the length of the arc,
- if E is an area, then $\mu(E) = S(E)$ is the exposed surface.

For the population exposed, ρ will be :

- for a site, the number of people allocated to that site,
- for an arc of the network, the concentration of vehicles per km on that section, multiplied by the average number of occupants per vehicle,
- for an area, the population density by surface unit. This density varies with the nature of the area exposed.

III - Functions of the Transportation risk assessment decision support system

This software includes five functions integrated into an interactive tool dedicated to the search for an optimal route (Figure 1):

- Definition of the problem of transport by entering or pointing on the screen : nature and volume of the substance, origine and destination, operationnal conditions.

- Definition of penalties on the network by marking prohibited nodes, links of the network, types of links of the network (eg., exclusion of road with only one lane) or areas,

- Generation of path by the Dijkstra algorithm with its graphical representation,

- Access to the data base to provide information about the links,
- Zooming into the map and loading of geographical overlays (ie.,hydrology, railways, communes).

COMPUTER ASSISTED RISK   285

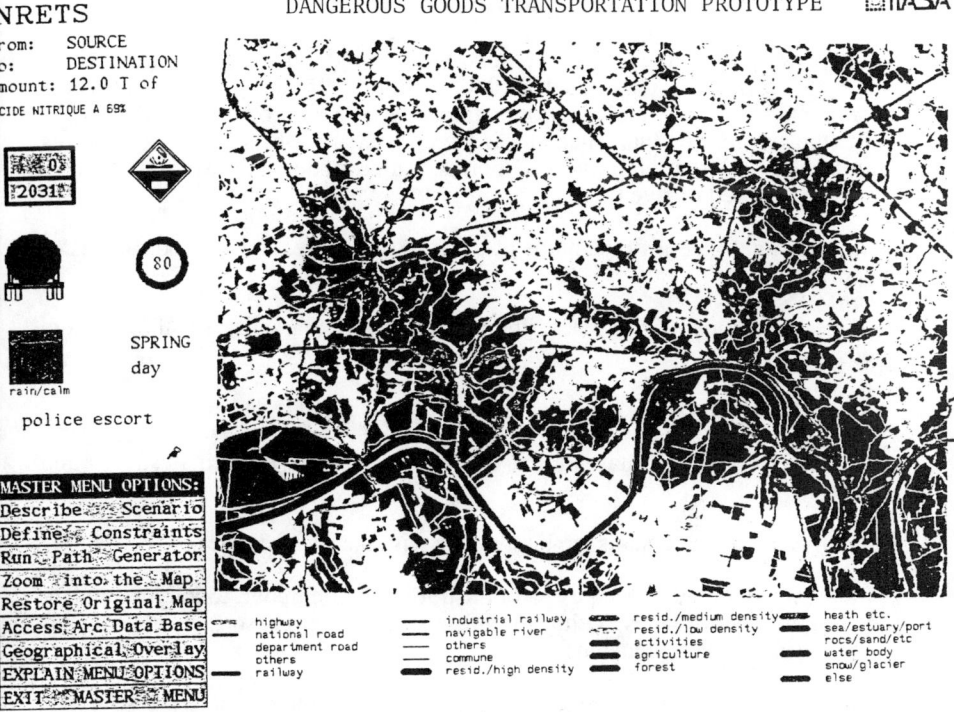

Figure 1: Basic screen for an interactive search of the less risky route.

V - Data sources

Two kinds of data are needed as input :

spatial data, which could be represented as a map or layer of spatial entities under a vector or raster-format,

non-spatial data, mostly statistical, which provides the values of attributes of spatial entities.

The aim is to create a general data structure able to handle these two kinds of data, which can be issued from different sources under different formats, and to connect a geographical object and a statistical object which is essential for the risk evaluation.

IV.1 - Spatial data

Most of the digitized cartographic data is draw from the BDCarto, the national data bank of digitized maps, developed by IGN (IGN, 1988, 1989b). The digitized information on the network and land use covers five maps on a scale of 1:50,000. More precise locations for special parts of the infrastructure, and industrial sites, will be digitized on nine 1/25,000 scale maps.

IV.1.1 - Transportation Network

The network for the three transportation modes : road, rail and water is digitized in a vector format by IGN at a 1:50,000 resolution. In fact, IGN provides two layers : one for road and rail networks, the other for hydrology (IGN, 1989a). Under the assumption that a network is a graph, an arc consists of a set of points from which the beginning and end points, called nodes, are differentiated. A layer is composed of five files :

- a header file, to identify the name of the map, the scale, and the coordinates of the corners,

- a code file, to identify and describe the codes and characters for the arcs and nodes,

- an arc file, which contains the list of arcs with their nodes and attributes,

- a node file, which contains a list of nodes with their coordinates, arcs and attributes,

- a point file, which contains the least points with their coordinates.

For arcs, there are ten attributes : administrative road number, toponym number, function, number of lanes, width, administrative categories, surface, node levels, ground position, miscellaneous.

For nodes, there are two attributes : toponym number, and characteristics.

IV.1.2 - Commune and land use

The map for communes is in a vector format as for the network, with a specification for the topology (code for area on right and on left of the arc), plus a special file for the areas with their centers and codes. Each commune is identified by an INSEE number.

Another map layer, which shows different types of land use, is issued by IGN (IGN, 1989b) with five main categories :

1. Built-up areas (5 ha. minimum surface)

   (a) Predominantly residential
       i. high density of residences (80 % - 100 % of buildings)
       ii. medium density (40 % - 80 % of buildings)
       iii. low density (10 % - 40 % of buildings)

   (b) Predominantly non-residential
       i. offices and industries
       ii. infrastructure for education, sport, health and social activities
       iii. large infrastructure for transport
       iv. quarries, open space in urban areas

2. Agricultural areas

   (a) cultivated land, (b) meadows, (c) vineyards, (d) orchards

3. Natural or low density built-up areas (8 ha. minimum surface)

   (a) woods, (b) moors, (c) mineral formations

4. Water bodies and related areas

   (a) open seas, tidal and non-tidal, (b) estuaries, lagoons (surfaces > 1 ha), (c) inland waters (non-perennial), such as swamps, salt-marshes etc., (d) inland waters (perennial) such as rivers (surface > 50 meters) lakes and ports, (e) beaches : rocky, pebble and sand

5. snow, glaciers, ice and permafrost

IV.1.3 - Infrastructure

The wide categories of land use mentioned above can be broken down further, especially for non-residential areas. A complete exposure to the risk must take account of the functions or activities of the non-residential areas, such as :

- Industry, services and commerce,
- Education, health and social activities, sport and cultural activities,
- Transportation.

These activities take place in buildings and related infrastructure, represented by areas or sites, according to the resolution of the map and the surface covered by these activities. In fact an area has to be decomposed into sub-areas, for example, an industrial zone is split up between different sub-zones owned by different firms, or contains different industrial sites.

The presence on the map is associated with statistical information on the activity area or site. Each fixed installation of the infrastructure is described by two attributes : permanent population and reception capacity. For a school, the labor force is reckoned on the one hand, and the number of scholars on the other.

For this feasibility study, importance is placed on the more vulnerable zones of sites in case of an accident, which concentrate a large amount of people difficult to protect or evacuate, eg. schools. Later, hospitals, commercial centers, camping grounds, etc., could be introduced.

All the sites of the primary, secondary or specialised schools in the zone have been digitized on maps of 1/25 000 scale.

IV.2 - Statistical data

The problem is no more the location on a map of a unit useful for the risk analysis, but the availability of statistical information pertinent to establish the exposure to risk according to the activity of the zone, or to estimate the volume of dangerous sustances transported in the network.

IV.2.1 - Population

The commune is the main statistical and geographical unit, taking into account the existing statistical data known about the network and its surroundings, available from the 1982 population census and commune inventory.

As a first step, the possibility of extracting commune by commune data from national statistical data banks have been examined (Lassarre, 1988), such as :

1. 1982 population census (INSEE),
2. SIRENE for firms (INSEE),
3. 1988 inventory for commune (INSEE).

The main source of information on residential dwelling is the 1982 census BDCOM82 file (INSEE, 1984a ; INSEE, 1984b) which provides, by communes, the number of persons for :

1. PT82 : the total 'legal" population, which is the sum of the total municipal population and the *comptée à part* category [1] of the population

2. Q82 : the total municipal population, which is the sum of the population of the principal residences, the population of mobile residences, and the population of collective communities.

3. R82 : agglomerated population (in the vicinity of the town hall) gives the population of principal residences and collective communities in the agglomerated part of the commune.

4. S82 : sparse population, which is the total municipal population, non-agglomerated.

From the INSEE data of the 1982 census, the distribution by age (1-20 years) of the population by sex and of the school population are available in table E041 for persons under 15 years, and in table E042 for persons between 15-20 years. In table E08, the population of pupils and students is provided in total, by sex and by class of age (15-19 years, 20-24 years, 25-29 years).

Statistics on active, employed population above 15 years of age extracted from the 1982 1/4 Population Survey, which can be disaggegated for communes with more than 8,000 inhabitants and aggregated for others, may also be used. INSEE Tables T12A and T12B distributes the active employed population into 15 and 40 economic activities for the three main communes : Yvetot, Lillebonne and Notre Dame de Gravenchon and the set of 39 other communes, with a distinction made between the workplace and the residence.

---

[1] As set out in INSEE, 1988, and which includes military forces, resident students, inmates of prisons, asylums, sanatoriums, etc.

1. Agriculture
2. Agricultural industries
3. Energy distribution and production
4. Intermediate goods production
5. Equipment production
6. Consumer goods production
7. Construction
8. Commerce
9. Transportation and telecommunication
10. Market services (charged)
11. Real estate
12. Insurance
13. Financial organisations
14. Non market services.

Table 2: List of economic activities for the nomenclature '15'.

4.2.2 - Flows

Two surveys were conducted in 1988 : one by questionnaire to the firms about their receipts and deliveries of dangerous goods, the other on the road on eight important junctions with a view to collect an O-D traffic-flow matrix.

V - Architecture of the system

The system is composed of four main modules.

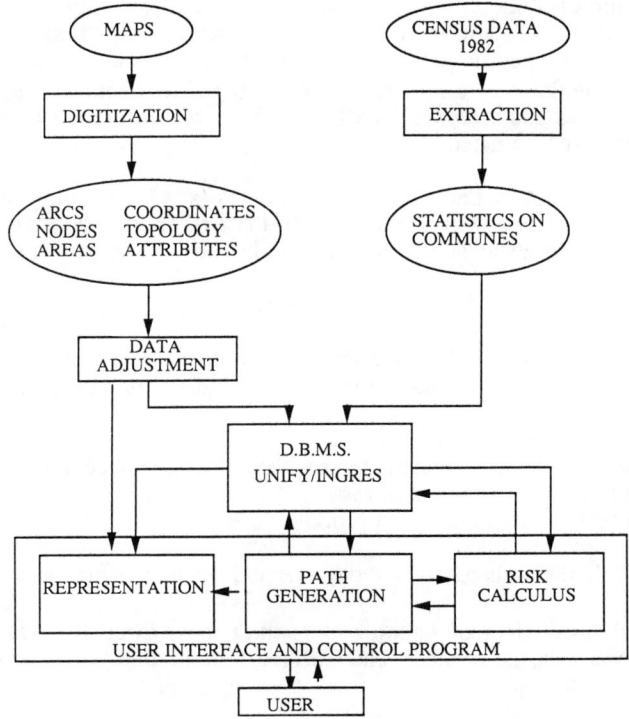

Figure 2: Schematic representation of the architecture of the system.

V.1 - A data manager

It is composed of different auxiliary modules and of a DBMS in order to create a general data structure able to handle complex geographical and statistical data, and to transform them into four patterns of data more easily accessible by the system :

- spatial information in vector format in the DLG standard (USGS, 1986) for the network of roads regularly maintained and the land use simplified into twelve categories to be available on the screen in the form of a map ;

- spatial information in raster format for data on railway network, hydrology and administrative boundaries to be displayed as an overlay on top of the basic map ;

- relational data base : the integration per se is here. Using the INGRES DBMS, a data base relating the different data types has been implemented. First each arc, area and node provided in digitized form is introduced into the data base and provided with additional information (eg. number of lanes, road type, or name and INSEE number of a commune) on its properties. Then all the statistical information is added to the data base and related to the areas, arcs and nodes (eg. population data to the communes). Third the area affected by a possible accident on a specific arc has to be provided : the surface affected (set as a stopgap solution to 1 km distance along both sides of the arc) is split into m^2 types of land use affected. The INRETS extension includes each commune where the arc is situated.

- binary files : for fast access, the coordinates (x, y) describing the track of all arcs, and the labels with the address of the first (x,y) coordinate for a section are stored, plus a special file with arcs and nodes for the path generator.

V.2 - The risk calculation module

The risk for an arc is calculated as the product of the rate of accident times the number of people affected in the accident and situated on three geographical objects :

- the territory defined as a parcel of specified land use on a commune,
- the link defined as a section of road,
- the site defined in our case as a school.

The risk calculation is based on a daytime and weekday situation.

The number of affected[1] people on each or these three objects is equal to the product of a measure of the spread (surface, length, unity) by a density.

---

[1] The impact zone is defined as a polygon constructed symmetrically on the axis of the arc, the width of which would depend on the category of the dangerous material transported. For the moment, the

For a link, we use a table giving the concentration in vehicles/kilometer by administrative category of road and an occupational rate of 1.8 occupant per vehicle.

	Concentration
Motorways	40
National roads	20
Secondary roads	5
Others	1

Table 3: Concentration by administrative category of roads

The problem is to distribute the statistics about the population within a commune according to land use. The population, by commune, is classified into :

- pupils or students

- those actively employed in :
    - agriculture,
    - industry,
    - services

- inactive

The population of pupils by commune is calculated by adding the sum over the classes of age from 1-14 of the school population in table E041 to the pupils and students of table E08 (> 15 years of age).

The population of inactive persons is calculated by adding the difference between the total population over the classes of age from 1-14 of the school population in table E041 and deducting the employed and unemployed active population in table E08 from the municipal legal population. This inactive population includes the real inactive people, plus young men during their military service, apprentices and unemployed youth who have not had their first jobs.

The active employed population is divided by commune (three main communes and one "super" commune) into

- agriculture (= 1),
- industry (= 2 to 7),
- services (= 8 to 14)

of the economic activities shown in table T12at (Table 2).

This population is distributed between the areas situated in the commune according to the following rules. The density is set to zero except for some land uses and schools.

---

maximum distance of impact from the axis is unique, and fixed to 1000 meters.

land use/population	pupils	inactive	agriculture	industry	services
high urban density	No	Yes	No	No	Yes
medium urban density	No	Yes	No	No	Yes
low urban dnsity	No	Yes	No	No	No
industries	No	No	No	Yes	No
agricultural area	No	No	Yes	No	No
schools	Yes	No	No	No	No

Table 4: Indicator of the distribution of population according to land use.

With some hypotheses on the ratio of densities between land use, it is possible to compute a density for each parcel of land use for the complete sample of communes. For others communes, a default value is attributed. The scholars of diffrent communes are distributed among the schools according to a gravity matrix. Once, those densities calculated are fixed, the consequences to the impact zone along an arc can be estimated and entered.

V.3 - The path generator

After the user selects a source and a destination, the path generator can be run to find the shortest path from source to destination. Djikstra's shortest path algorithm has been used here (Boffey, 1982), with the risk value taken as the criteria for optimization.

It computes the path with the lowest risk between a source and a destination, with integration, if required, of constraints on the network.

V.4 - The user interface and control program

It handles the menu-driven dialog with the user via the mouse, as well as all the display and the integration of the above modules into one coherent system.

VI - Implementation

The software developments use a UNIX-based open architecture ; the system is highly modular and is open to further extensions and improvements.

The custom-configured software is written in C, using the host language interface of INGRES and has been implemented on dedicated, high-resolution color graphics workstations (SUN).

References

Abkowitz, M. and Cheng, P. (1988) Developing a Risk/Cost Framework for Routing Truck Movements of Hazardous Materials. Accident Analysis and Prevention, 20(1), 39-51.

Boffey, T.B. (1982) Graph Theory in Operations Research. Macmillan Press Ltd.

Fedra, K., Lassarre, S., Weigkricht, E., (1989) Evaluation du risque de transport de matières dangereuses assistée par ordinateur sur la Haute-Normandie. Convention d'étude DTT/INRETS. Final report by International Institute for Applied Systerms Analysis, Laxenburg, Austria.

Hubert, P., Pagès, P., and Degrande, J.P. (1986) L'évaluation du risque d'accidents graves dus au transport de matières dangereuses dans la région sud de Lyon. Rapport 95. Centre d'Etude sur l'Evaluation de la Protection dans le domaine Nucléaire. Fontenay aux Roses. France.

Hubert, P., Pagès, P. (1989) Risk managemnt for hazardous materials transportation: a local study in Lyons. Risk Analysis, 9(4), 445-451.

IGN (1988) Descriptif du format des fichiers BDcarto. Service des cartes et dérivés thématiques. Département base de données cartographiques. Saint-Mandé.

IGN (1989a) Base de données cartographiques spécifications du contenu. Service des cartes et dérivés thématiques. Département base de données cartographiques. Saint-Mandé.

IGN (1989b) Description du format des fichiers BDcarto pour l'occupation des sols. Service des cartes et dérivés thématiques. Département base de données cartographiques. Saint-Mandé.

IIASA (1988) Annual report 88, International Institute for Applied Systerms Analysis, Laxenburg, Austria.

INSEE (1984a) Recensement de la population 1982 - Description de la bande de données communales. INSEE. Paris.

INSEE (1984b) Recensement général de la population de 1982. Guide d'utilisation Tome 1. INSEE Paris.

INSEE (1988) Recensement général de la population de 1982. Census of the Population of 1982. Annuaire Statistique de la France 1988. Résultats de 1987. Vol. 93. Institut de la statistique et des études économiques. Paris, France.

Kessler, D. (1987) Establishing Hazardous Materials Truck Routes for Shipments through the Dallas Fort Worth Area. Transportation Research Board. State of Art Report 3,379-87.

Lassarre, S. (1988) Logiciel d'évaluation du risque du transport de matières dangereuses. INRETS-DERA. Arcueil.

OECD (1988) Transporting Hazardous Goods by Road. OECD, Paris.

Pijawka, K.D., Foote, S., ans Soesilo, A (1985) Risk Assessment of Transporting Hazardous Materials : Route Analysis and Hazard Management. Transportation Research Record 1020, 1-6.

Saccamanno, F.F. and Chan, A.Y.W. (1985) Economic Evaluation of Routing Strategies for Hazardous Road Shipments. Transportation Research Record 1020, 12-18.

Saccamanno, F.F., Shortreed, J.H., Van Aerde, M. and Higgs, J. (1989) Comparison of Risk Measures for the Transport of Dangerous Commodities by truck and Rail. Paper N° 88-0520. Paper presented the 68th Annual Meeting of the Transportation Research Board. Washington, D.C. January 22-26, 1989. Institute for Risk Research. University of Waterloo. Waterloo, Canada.

USGS (1986) Digital Line Graphs from 1:24,000-Scale Maps. Data Users Guide 1. United States Department of the Interior U.S. Geological Survey, Reston, Virginia.

Van Aerde, M., Shortreed, J., Lind, N., McBean, B., Needleman, L., Saccamanno, F.F. and Yagar, S. (1987) Risk Management in the Handling in Transportation of Dangerous Goods. Phase II Final Report. Vol. 1. Appendices Vol. 2. Institue for Risk Research. University of Waterloo Research Institute.

DATE DUE